Nurhon Isaeva

# CATALYSTS FOR HYDROTREATMENT OF OIL FRACTIONS

Nurhon Isaeva

# CATALYSTS FOR HYDROTREATMENT OF OIL FRACTIONS

## Overview and new developments

ScienciaScripts

**Imprint**



Cover image: www.ingimage.com

This book is a translation from the original published under ISBN 978-3-659-95847-2.

Publisher:
Sciencia Scripts
is a trademark of
Dodo Books Indian Ocean Ltd. and OmniScriptum S.R.L publishing group

120 High Road, East Finchley, London, N2 9ED, United Kingdom
Str. Armeneasca 28/1, office 1, Chisinau MD-2012, Republic of Moldova, Europe
Managing Directors: Ieva Konstantinova, Victoria Ursu
info@omniscriptum.com

Printed at: see last page
**ISBN: 978-620-8-40643-1**

**Nurkhon ISAEVA**

# CATALYSTS FOR HYDROTREATMENT OF OIL FRACTIONS

(overview and new developments)

*Monograph*

Tashkent - 2024

**UDC** 541.183:661.183.2

**Nurkhon Farhatovna Isayeva.**

Catalysts of hydrotreating of oil fractions [Text] / Isaeva N.F. -2024. - 160 c.

The book presents the production of oil hydrotreating catalyst, the method of synthesis of wide-porous carrier, physicochemical and other properties of oil hydrotreating catalyst, production and innovative technological processes developed on the basis of physicochemical research. Regularities and theoretical dependencies of interaction of cobalt, nickel and molybdenum salts, methods of their application on the carrier, methods of thermal treatment are considered, technology of obtaining trimetallic cobalt-nickel-molybdenum catalyst for hydrotreatment of oils is proposed.

The book is intended for scientific and engineering technicians of scientific and design centers, as well as for masters of universities, doctoral students in the direction of catalysis and petroleum refining.

**UDC 541.183:661.183.2**

**Reviewer:**

**M.U.Karimov -** *Professor, Deputy Director for Science, Chemical and Technological Research Institute*

Tashkent, 2024

# Table of contents

# INTRODUCTION

Increasing the production of petroleum products, expanding their assortment and improving their quality are the main tasks set before the oil refining industry at present. The solution of these tasks in conditions when the share of processing of high-sulfur and high-paraffin crude oils is continuously increasing, more and more heavy fractions are involved in the refining process, allowing to increase the depth of oil refining in order to increase the yield of fuels with improvement of their quality, required changes in oil refining technology and stimulated the increase in the capacity of hydrogenation processes of oil refining, primarily hydrotreating, hydrotreating and hydrocracking, which are expected to continue in the near future. In this connection, in order to meet environmental requirements, the issues of deep hydrotreating of oil fractions through the use of new modern catalysts and improvement of technological processes have become more and more acute in recent years.

The Fergana Refinery is the only enterprise in Central Asia producing more than 20 types of oils for various purposes, some of which are exported. Currently, motor oils produced by this enterprise comply with the ARI (American Petroleum Institute) classification group I, i.e. sulfur content within 0.5%, saturated hydrocarbons less than 90%, etc.

At the same time, the main volume of imported oils produced corresponds to ARI group II, where the content of sulfur and saturated hydrocarbons should be <0.03 and >90% weight, respectively.

In the world today, the volume of refined oil is more than 4.2 billion tons per year. The volume of hydrotreating processes accounts for 46% of the total volume of oil refining, so the demand for hydrotreating catalysts in the world is growing from year to year[1]. In this connection, one of the most urgent problems is the development of technology for obtaining new, more effective cobaltnickelmolybdenum catalysts for hydrotreatment of petroleum products and the study of their performance characteristics.

At present, the main task set before the oil refining industry is to meet the demand for petroleum products, expand their range and improve their quality. Improving the performance of petroleum products is mainly carried out with the help of hydrotreating process, therefore, scientific works aimed at developing new technologies for obtaining hydrodesulfurization catalysts are carried out.

During the years of independence in the Republic of Uzbekistan oil refining industry has achieved certain successes in the reconstruction of production, improvement of technological processes with the production of new products. However, insufficient attention is paid to development of domestic catalysts on the basis of local raw materials. One of the tasks reflected in the documents on the strategy of action for further development of the Republic of Uzbekistan is "Creation of technologies for import-substituting products from local raw materials and secondary resources"[2]. Organization of production of domestic hydrotreating catalyst will allow to replace imported catalyst and save significant foreign currency funds.

In this regard, the present work, aimed at developing a new domestic technology for the production of oil hydrotreating catalyst, is highly relevant.

Scientific research on the development of technology for obtaining catalysts for hydrotreating of oil fractions was carried out by such scientists as Startsev A.N., Nefedov B.K., Kogan L.O., Tomina N.N., Pimerzin A.A., Pashigreva A.V., Klimov O.V., Sidelkovskaya V.G., Eremina Yu, Pashigreva A.V., Klimov O.V., Sidelkovskaya V.G., Eremina Y.V., Surin S.A., Aliev R.R., Chukin G.D., BianchiniC., Topsoe H., Hughes R, ZhouL., Kaluza L., ZdraziL.M., Sultanov A.S., Yunusov M.P., Turabzhanov S.M., Saidakhmedov S.M. and others.

They have developed theoretical bases for synthesis of catalysts for hydrotreatment of gasoline, diesel fuels, oils and protective layers for these processes. They gave recommendations on implementation of modern technologies for production of active catalysts with high economic and technological properties. In particular, a new direction of synthesis of catalysts for hydroprocesses without a carrier, showing increased activity due to high dispersibility and concentration of active catalyst components, are recommended for production.

At the same time, scientific research is underway to study the formation of active catalyst structures, taking into account the specifics of the processed feedstock, as well as to develop highly efficient technologies for the preparation of effective catalysts for hydrotreatment of petroleum products, at low energy costs and losses of raw materials

# CHAPTER I. CATALYSTS FOR HYDROTREATING OF OIL FRACTIONS.

## §1.1 Prospects for development of hydrogenation processes in the oil refining industry of Uzbekistan

Catalytic hydrogenation processes are the basis for the production of many commercial petroleum products [1; P.26-27]. The urgent problems of hydroprocesses in the world practice are usually solved by construction of new plants operating at high hydrogen pressure. The oil refining industry of Uzbekistan is also faced with the task of deepening the processing of oil through the intensification of production, reconstruction of existing facilities, expansion of the range and quality of petroleum products. An innovation in the development of local high-sulfur feedstock at the Fergana refinery in the production of oil oils with viscosity index above 100 was a change in the mode of hydrotreating of oils by reducing the operating pressure to 2.5 MPa from 3.8 MPa, typical for feedstock from West Siberian oil, achieved due to the successfully selected and justified ratio of oil: gas condensate [2; P.39]. Constant increase of requirements to technical oils and the tendency to tightening of these indicators, make us continuously optimize the production technology for obtaining high quality oils on the basis of available natural resources, because high quality base oils are the basis for a promising range of products [3; P.22-23]. Many countries, including Russia [4; P.34], have already introduced new standards for quality indicators of base mineral oils according to API classification [5; P.15]. API classification divides motor oils into two categories: "S" - oils for gasoline engines and "C" - oils for diesel engines [6; P.138-139]. Base oils, depending on the viscosity index, degree of saturation of the content of paraffinonaphthenic hydrocarbons and sulfur are subdivided into 5 groups. New standards and increasing competition in the world market, first of all, stimulate the development of new technologies that purposefully change the chemical composition of base oils. In this respect the processes of dewaxing and iso-dewaxing are promising [7; .124, 8; P.23, 9; P.59-60, 10; .7-8], which are comprehensively investigated [11; P.26-28, 12; P.8-10]. In modern world oil refining the most urgent and complicated problem is refining (demetallization, deasphalting and desulfurization) and catalytic processing of oil residues [13; P.579], in general for the last 20-25 years the world production of lubricating oils on the basis of catalytic hydrogenation processes, including hydrotreating and catalytic dewaxing and hydroisomerization [14; P.17], has increased more than three times. According to the Internet forecast, Roadmap "Use of nanotechnologies in catalytic processes of oil refining". From 30.10.2010. the role of hydroprocesses in obtaining various petroleum products will only increase. The share of hydroprocesses in the technological structure of production in the USA and Japan is 26 and 29.5%, respectively. In Russia 96.5% of base oils are obtained by extraction purification and dewaxing with selective solvents. Purification of oil fractions by selective solvents is the main process of solvent technology of petroleum oil production. It is intended for removal of resinous substances, polyaromatic hydrocarbons, naphthenoaromatic hydrocarbons with short side chains, sulfur-containing and organometallic compounds from oil distillates and deasphalticates. In Uzbekistan phenol is used as a solvent for selective purification. Furfurol and N-methylpyrrolidone are becoming more widespread [15; P.150]. In spite of the development of new generation catalysts, improvement of advanced oil refining processes, including the involvement of secondary diesel fractions in the feedstock [16; P.3-4, 17; P.124-125, 17; P.4, 18; P.16, 19; P. 4-6, 20; P.18-20, 21; P. 26-38; 22; C. 199], up to now typical G-24 and G-24/1 units of many Russian refineries are not equipped with modern hydrotreating reactors

**Figure 1.1. Global market of catalysts for processes with the highest market prospects**

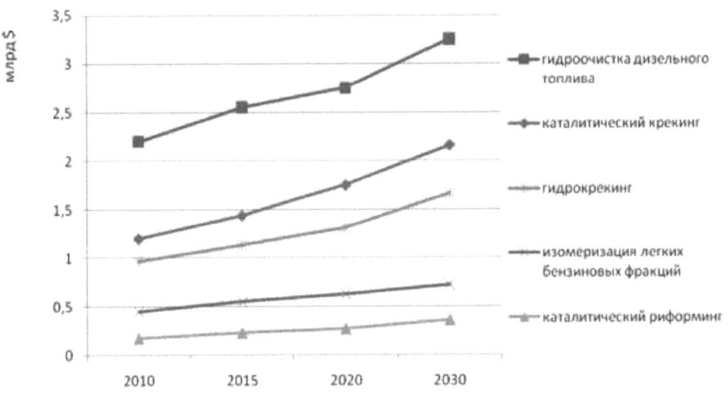


A similar situation has developed in Uzbekistan [23; P.3-5, 24; P.71-73], although the total economic efficiency from the organization of production of oil oils and lubricants at the Fergana Oil Refinery, where the main component is oils, for the period 1994-2002, amounted to 110 million soums [2; P.6]. [2; C.6]. The material base of the process of hydrotreating of heavy oil fractions - raw materials for the production of base oils, is represented by outdated reactors designed for low hydrogen pressure. At Fergana oil refinery mineral oils are produced according to the traditional scheme: deasphalting (residual fraction - tar sands) [25; P.15], phenol purification, dewaxing and hydrotreating. Hydrotreating on a stationary bed of aluminocobalt-molybdenum (ACM) catalyst GO-70 is subjected mainly to the third fraction of vacuum distillate and desphalting tar. In order to ensure the required properties of specific grades, base oils are prepared by compounding of hydrogenizates. Hydrotreating improves the quality of oils by main indicators: viscosity index by 1-2 points, color by 2-4ed. CNT, decreases coking by 0,05-0,15% and sulfur content to 0,3-0,4% [26; C.15]. The degree of hydrogenation of organosulfur substances depends on the processed raw materials and is, as a rule, 30-40% [27; P.84]. Application of this technology makes it possible to obtain base oils that meet the requirements for Group I oils according to API classification. If the quality of the base oil does not meet the standard, the composition of the additive composition has to be changed in order to obtain commercial oils of the required grades [28; C.37-39, 29; C.36, 30; C.23-24, 31; C.1, 32; C.33]. Group II base oils in the countries of Western Europe are produced in the process of hydrocracking, which allows to obtain low content of sulfur and aromatic compounds. Production of group II oils in Russia is based on a combination of hydrocracking and selective purification processes. Group II, III oils should not contain resins, aromatic hydrocarbons content should not exceed 1.5 wt. %, and sulfur content should be less than 300 ppm.

Domestic base oils are significantly inferior to foreign analogues in viscosity index by 20-40 points, in sulfur content - by dozens of times and significantly - in color and its stability and storage process. To some extent phenolic purification helps to improve the situation, but at the same time, as it is known, the yield of raffinades decreases [28; P.38-39]. It is possible to solve this problem radically only by introducing hydrocracking or by combining the processes of partial dewaxing, selective hydrocracking and hydrotreating [33; P.27]. The hydrocracking process proceeds at pressure higher than 10 MPa and involves into production undesirable hydrocarbon and non-hydrocarbon components of processed dewaxed oils. Technological equipment for oil production in Uzbekistan is not designed for operation at pressure of 10-15 MPa, and domestic catalysts for hydrocracking under milder conditions are not available. It should be noted that most of the studies conducted in Uzbekistan, as well as in other countries, are aimed at increasing the yield of light oil products [34; P.7-11, 35; P.30-32, 36; P.63, 37; P. 51-52, 38; 44-45, 39; C.28, 40; C.28-29, 41; C.33 42; C.26, 43; C.15-16, 44; C.16], as the most demanded large-capacity commodity products. Considerably less works [29; C.34, 30; C.23,45; C.141, 46; C.11, 47; C.27-28, 48; C.18-19, 49; C.12-13], including the development of catalytic protective systems [50; C.55, 51; C.1, 52; C.1, 53; C.1, 54; C.48-49], are devoted to the issues of improvement of technical and economic indicators of commercial oils production.

Thus, in order to obtain commercial oils that meet international standards, it is necessary to increase the depth of removal of undesirable components by selective purification and the subsequent stage of hydrotreatment of vacuum distillates and deasphaltized residue. This will allow to refuse import of motor oils and lubricants, satisfying the demand of our country's consumers with domestic products.

## §1.2 Ways to improve the quality of base oils

Processing of heavy oil fractions is determined by specific conditions: chemical composition of hydrocarbon raw materials [27; P. 84], nomenclature of required products and technological capabilities of equipment. The Republic of Uzbekistan has reserves of heavy sulfur oils [55; P.63-64]. Analysis of the main components of oil raw materials produced in Uzbekistan [56; P.63-64, 57; P.59-60] shows a significant difference in the viscosity characteristics of crude oil, as well as the content of paraffins, resinous asphaltene substances and sulfur. The ratio of these components and group hydrocarbon composition largely determines the choice of refining schemes taking into account the needs of the petroleum product market. Refining of heavy feedstock, oriented at increasing the yield of fuel fractions, requires catalysts capable of active cleavage of resinous asphaltene substances into light petroleum products. Therefore, there is a constant search for ways to intensify the processes aimed at increasing the yield and quality of fuel fractions [58; P.21-23, 59; P.13, 60; P.25]. If the target product is oils, the quality problem can also be solved by hydrogenation processes. Their principal difference from all other processes of oil production consists in providing the required oil quality not by removing low-value or harmful components, but by their chemical transformation. As a result of deep hydrogenation of aromatic hydrocarbons and partial opening of naphthenic rings paraffin-naphthenic hydrocarbons become the main components of oil fractions. At the same time, the fractions obtained by hydrocracking can be directed to oil production, bypassing the stage of selective purification [61; P.58-60].

Recently, a number of new catalytic processes have been introduced abroad that allow changing physical and operational properties of petroleum products in the presence of hydrogen: catalytic dewaxing, hydro-isomerization, hydroalcoholization, hydrodealkylation and others [62; P.21, 63; P.3, 64; P.36-37]. Their realization makes it possible to obtain the necessary target products with high yield practically from any oil feedstock. Naturally, for this purpose it is necessary to create new specific catalysts effective at low temperatures and pressures.

When obtaining oils, the decisive properties of catalysts are decolorizing ability and viscosity index increase, with high resistance to heavy metals. To choose the most rational technology of oil distillate purification [65; C.36, 66; C.22], which provides obtaining oils with specified properties and maximum yield, it is necessary to have a sufficiently complete idea of the chemical composition of raw materials and properties of individual groups of hydrocarbons included in it. The most important operational characteristics of base oils depend on their chemical composition: viscosity, viscosity index, pour point and others. It has long been known that viscosity index of hydrocarbons changes in the following series: n-alkanes> isoparaffins with single branches> isoparaffins with multiple branches ≈ mononaphthenes with long side chainsmonocyclic aromatics with long side chains> polycyclic aromatics ≈ polycyclic naphthenes with multiple attached short chains. Thus, the quality of feedstock is largely determined by the ratio of high-index and low-index polycyclic aromatic hydrocarbons.

Among the undesirable components requiring mandatory removal are residual asphaltenes, resins, sulfur and nitrogen compounds [61; P. 15-21]. The ratio of the named classes of substances depends, first of all, on the properties of crude oil, and secondly, on the scheme of its processing. It is economically and technically expedient to refine for oils those oils, in heavy fractions of which "desirable" components prevail. A large amount of resinous asphaltene substances, polycyclic aromatic hydrocarbons, sulfur-containing and other hetero compounds complicates processing, contributes to low yields of target products and in many cases does not allow to ensure the desired quality [61; P.10].

The process conditions, namely temperature, pressure, feedstock feed rate and hydrogen-containing gas circulation have a decisive influence on the depth of conversion of hydrocarbon feedstock and, consequently, on the yield and quality of the target products. Hydrotreating of vacuum gas oil boiling up to 500°C is usually not very difficult. At moderate hydrogen pressure of 4-5 MPa, temperature of 360-410°C and feedstock feed rate of 1.0-1.5 $h^{-1}$ the degree of desulfurization of oil fractions usually reaches 89-94%.

At the same time the nitrogen content decreases by 20-35%, metals - by 75-85%, aromatic hydrocarbons - by 10-12%. At pressure of 3.0- 3.5 MPa these indices are naturally lower. [66; 128-129].

A certain effect for improvement of color characteristics is given by narrowing the fractional composition of oil distillates when using a two-column purification scheme with phenol and furfurol [67; C.22-23, 68; C. 10]. The replacement of propane, in the process of tar deasphalting, with a weighted propane-butane mixture allows to vary the characteristics of deasphalticate more flexibly to obtain a high-viscosity base oil [25; C. 14].

It is established [28; P.37-39] that the required level of quality of the commercial product is achieved only when using base oil of optimal chemical composition and balanced composition of additives of functional action. The main condition for intensification of oil production is constant composition of raw materials. Without observance of this condition neither change of technological processes regimes nor use of effective solvents will allow to obtain oils of required quality. The oil obtained from raw materials of worse chemical composition contains less high index components and more resins. This is fully applicable to the Fergana Refinery, which is characterized by strong fluctuations in the chemical composition of raw materials. As a consequence, base oils differ significantly in their properties and do not always meet the criteria of international standards [28; P. 38]. The use of traditional additives introduced in this base did not give the expected results. And only the use of additive package Lubrizol-4979 with zinc dithiophosphate and Lubrizol-1395 in combination with the additive Ferad, allowed to obtain a motor oil close in quality to the oil prepared on a better purified base. Such a way is justified when the variability of raw material composition and insufficient content of hydrocarbons desirable for oils limit the possibility of obtaining a base of satisfactory quality.

Thanks to the complex of works carried out at the Fergana Oil Refinery, the shortage of oils that had arisen when foreign analogs were too expensive was eliminated. The production of turbine, spindle, transformer, hydraulic and other types of oils has been established in the territory of the Republic of Uzbekistan. However, unstable quality of raw materials, wear of equipment and use of outdated catalysts, operational and physical-chemical properties of base oils do not always allow to meet the requirements of modern consumers.

Development of new generation hydrotreating catalysts capable to significantly improve the main characteristics of base oils at hydrogen pressure of 2.5-3.0 MPa is a way to solve this urgent problem without significant reconstruction of the existing hydrotreating unit.

### §1.3 Synthesis of hydrotreating catalyst carriers

Numerous studies have long ago established the basic requirements for the texture of aluminum hydroxide-based carriers for hydrotreating catalysts for various petroleum products. Specific parameters (pore size, pore radius distribution, pore shape) are determined by the peculiarities of the catalytic process. In particular, the processing of oil fractions requires the presence of a significant number of large pores, since the size of the processed molecules can reach several nanometers. The pore diameter should be not less than the size of the molecules present in the oil so that the progression of reagents to the active component of the catalyst is not hindered [69; P. 155]. Taking into account economic indicators, the possibility to flexibly adjust the texture and maximum reactivity of freshly precipitated aluminum hydroxides, catalyst manufacturers try to organize the process chain in such a way as to bypass the stage of drying of aluminum hydroxide precipitate [70; P.53-54]. By combination of hydroxides of "hot" and "cold" precipitation mechanically strong carriers with biporous structure necessary for synthesis of catalysts for hydrotreating of high boiling feedstock were created. The mentioned technology of coarse-porous carriers production has been mastered on an industrial scale at Russian enterprises (Angarsk plant of catalysts and organic synthesis). The role of particle sizes of freshly deposited and "aged" precipitates of a number of modifications of aluminum hydroxide in the formation of the optimal texture of carriers is considered in detail in the works by A.A. Lamberov, O.V. Levin et al. [71; P.54-63; 72; P.153, 73; P.305-307].

Fundamentally new methods of mechanochemical and thermochemical activation to produce phase-homogeneous aluminum hydroxides have been developed at the Institute of Catalysis, Siberian Branch of the Russian Academy of Sciences [74; P. 761-762]. The enormous influence of special additives [75; P.1, 76; P.1, 77, P.1, 78; P.1, 79; P.1] and impurities included in the composition of raw materials on the properties of carriers and catalysts is reflected in the scientific patent literature. Sidelkovskaya V.G. and co-authors [80; P.1417-1418] showed that the addition of 25-50% of Troshkov clay in the carrier is quite acceptable in the production of active aluminonickelmolybdenum (ANM) catalyst, so reduces the number of non-active spinel structures. It is proved that the formation of low-dispersed polymolybdate structures including nickel ions and nickel silicate formations reduce the activity of catalysts. The authors of the patent [81; P.1] limit the content of clay minerals (kaolin) in the catalyst to only 12 %. The cycle of works with Angren kaolin [82; P. 942, 83; P.28-30] led to the conclusion about its suitability in the synthesis of carriers for catalysts of hydrodesulfurization of diesel fuel and natural gas. Since clays are natural aluminosilicates, under certain conditions prone to zeolitization, it should be noted the works devoted to the study of the nature of interaction of nickel and molybdenum ions with carriers of different nature, including those containing silicon in the form of amorphous aluminosilicates [84; P.125, P.129, 85; P.57-58; 86; P.231-233] or zeolites. Silicate additives allow to obtain carriers containing micro- and meso-pores, in the absence of macropores. It is noticed that in zeolite-containing catalysts there is a more effective interaction of nickel and molybdenum with the formation of a large number of nickel-molybdenum associates of different stoichiometric composition. It is suggested that the acidic centers of zeolite are the localization sites of nickel-molybdenum associates.

The main component of carriers - aluminum oxide also has pronounced acidic properties [84; P. 112-113]. Any additives, such as boron compounds [87; P. 349-353, 88; P. 152-154, 89; P. 523, 90; C. 83-85, 91; C. 544-545, 92; C. 258-257], fluorine [93; P.298-299, 94; P.241], phosphorus [95; P. 137-140, 96; P.169], as well as other elements or their combinations, changing the spectrum of acid-base properties, certainly affect the process of catalyst preparation, shifting the pH of the impregnating solution in the contact zone with the carrier and can change the structure of active centers and catalytic properties [87; 349-353]. There is no consensus in the literature on the role of acidity of catalysts. From some literary sources it is known that on strong acid centers undesirable processes of coke deposition and cracking with breaking of C - C bonds occur more intensively, so preference is given to samples with moderate acidity. The authors [97; P.492-493, 98; P.198], on the contrary, consider that acid centers enhance the hydrodesulfurizing ability of the active phase - molybdenum sulfide component and the higher the contribution of "acidity" to the total hydrodesulfurizing activity, the lower the relative hydrogenating activity. In order to increase the acidity of the catalyst, the carrier was sometimes specially treated with boric acid or boron trifluoride etc. [70; PP. 60-64, 99; PP. 3-4, 100; P. 76]. The literature provides information on the changes in morphology and acid-base properties of the surface at various stages of synthesis of alumina and alumina-kaolin carriers with the use of various acids

Thus, when synthesizing carriers for oil hydrotreating catalysts, it is necessary to ensure, first of all, an optimal combination of micro- and meso-pores, as well as the prevalence of acid centers of moderate strength on the surface in the range of $pK_a$ from - 5 to - 3.

## §1.4 Influence of various factors on the formation of catalytically active structures

The activity and selectivity of the catalyst is determined primarily by its chemical and phase composition. At the same time, the phase composition depends not only on the nature and quantity of the introduced ingredients, but is largely determined by the method of preparation [101; P. 44]. The analysis of patent and scientific literature has shown that modern technologies of preparation of catalysts for hydroprocesses of oil products refining are quite diverse

[102; P. 4255-4257, 103; PP. 504-506, 104; PP. 61, 105; 1254-1256, 106; C. 283-284] and reflect the peculiarities of the methods developed by the firms to regulate these or those physicochemical and mechanical properties of carriers and catalysts. The expediency of application of this or that catalyst and the method of its production is determined by the technical and economic indicators of the process in which it is used, as well as by the level of technology development. At first, massive catalysts synthesized by precipitation with concentrated ammonia solution from a mixed solution of nickel nitrate and ammonium paramolybdate at a temperature of about 80° C followed by forming and calcination at 500° C for 4 hours were widely studied. Up to 80 % of all oxide catalysts, catalyst carriers and sorbents are obtained by co-precipitation methods [101, P.46]. The essential advantage of the co-precipitation method is the possibility to vary the porous structure within wide limits. The disadvantages of include a large consumption of reagents and a significant amount of wastewater, as well as the fact that the precipitated contact masses retain impurities in the form of acid residues and products of incomplete hydrolysis, which, in turn, can be catalytic poisons.

Then more active and economical aluminocobaltmolybdenum (ACM) and aluminonickelmolybdenum (ANM) catalysts on carriers were investigated and introduced into industry. In the works of Landau M.V. et al [107; P. 931-937] it is shown that during coprecipitation from aqueous solutions of nickel, molybdenum and aluminum salts in acidic medium, binary molybdates $Ni_xMo_yO_z$ are preferentially formed, $AlMo_yO_z$ the formation of double molybdate of nickel and aluminum

$Ni_{(x)}Al_yMoO_4$ or heteropolymolybdate of nickel $N_x[Al_yMo_zO_nH_m]$ is possible. In alkaline medium due to the amphoteric nature of hydrated aluminum ions, the latter are present as anions of the $[A1(OH)^{(}{}_{3+x)}]^{x-}$ type. Therefore, along with nickel molybdate, formation of nickel alumomolybdate of $Ni(MoO_4)_{(x*)}(Al_2O_{(4))(y)}$ type similar to $NiMoO_{(4)}$, in which part of $[MoO_{(4)}]^{2-}$ anions is isomorphically substituted for aluminate anions is possible. It was found that the formation of the precipitate at increasing pH is due to the formation of poorly soluble nickel molybdonickelate $Ni_x[Ni_{(7-)(x)}Mo_7O_y]$, as evidenced by the broadening of bands in Raman spectra as a result of the incorporation of Ni into polyanions of molybdenum, that is, the formation of nickel molybdonickelate [108; P.936]. It should be noted that the Raman spectroscopy method (especially in combination with IR and electronic spectra) is very informative for studying the degree of polymerization of molybdate anions both in solution and as part of solid catalysts [84; P. 141 - 153, 109; P. 263, 110; P. 1267].

The technology of catalysts by applying the active component to the carrier has a number of advantages over others: relative simplicity, less harmful waste, more efficient use of the active substance. According to the method of introduction of active components Ni, Co, Mo into the carrier, two fundamentally different technologies can be distinguished: impregnation of granules of the prepared carrier with metal salt solutions and mixing of metal salts with the wet mass of the carrier or its precursor with subsequent granulation of the resulting mass, while co-extrusion is an intermediate option. Coextrusion is used when it is necessary to introduce into the composition of the catalyst mass a large amount of difficult soluble compounds, in particular molybdenum [111; P. 179, 112; P. 186]

In turn, the impregnation method is also applied in different variants. Most researchers prefer to apply molybdenum salt first, either by impregnating $\gamma$-$Al_2O_3$ by moisture absorption [111; P. 179, 113; P. 153] in 20-60 minutes, or by keeping the carrier granules in excess of ammonium para-molybdate solution for a long time [114; P.456-457]. After appropriate heat treatment (calcination temperature may vary from 200°C to 600°C) the obtained intermediates are impregnated with nickel or cobalt nitrate solution and again subjected to drying and final calcination to obtain ANM or ACM catalysts, respectively. Some authors consider it advisable to deposit nickel first and then molybdenum after intermediate calcination to obtain the most active structures. In general, it is considered that impregnation catalysts are more active and stable than co-precipitated analogs. By comparing the methods of impregnation, mixing and co-precipitation of ANM catalysts, it is shown that the carrier in hydrotreating catalysts performs not only the role of a substrate on which active phases are formed, but also exhibits hydrodesulfurizing activity in the reaction of hydrogenolysis of sulfur compounds.

In an effort to simplify the preparation technology, the authors of the article [115; P. 503.] applied impregnation of carriers with ammonia solution of a mixture of salts of nickel nitrate and ammonium para-molybdate. Due to the formation of nickel ammoniacates and good solubility of ammonium para-molybdate in the basic medium, it was possible to deposit 4 wt. % NiO and 14 wt. % $MoO_3$ on the granules of alumina carrier with the addition of amorphous and crystalline aluminosilicates. Good results were obtained impregnation with a joint solution of cobalt (or nickel) salts and ammonium para-molybdate [116; P. 947] stabilized with nitric or phosphoric acid [117; P. 444-445, 118; P. 921]. Phosphoric acid in the composition of impregnation solutions, unlike nitric acid, is not removed during heat treatment of impregnated pellets, but remains in the composition of the finished catalyst and can affect the formation of active centers. The influence of the preparation method and phosphorus concentration [119; P.500-503, 120; P.105, P. 117-118, 121; P. 1423, 122; C. 374, P.380-381] as part of the oxide form of hydroprocess catalysts on their specific activity, including for the processing of crude deposit "Maya" [123; PP. 10942-10944, 124; PP. 34-35]. It has been experimentally proved that phosphorus in the composition of catalysts, depending on its concentration and method of introduction, not only affects the hydrodesulfurizing and hydrodenitrating function [125; P. 359-360], but also contributes to the selectivity of target processes [126; P. 863]. This applies not only to catalytic systems on commonly used carriers, but also to such as zeolites [127; P. 3924], carbides [128; P. 239] and nitrides [129; P. 204-205] .

The formation of complex compounds with transition metals, especially in combination with other inorganic [130; P. 81, 131; P. 136] and organic complex-formers [132; P. 173-175], such as urea [133; P. 904, P. 906-907], is considered to be one of the main reasons for the positive effect of phosphoric acid or its derivatives on the catalytic activity of diesel hydrotreating catalysts. Heteropolyacids [134; P. 251, P. 255] and their salts are considered as precursors of active phases. Most often in the synthesis of hydrotreating catalysts the carriers are impregnated with heteropoly compounds based on heteropolyacids of the 6-series with Anderson structure [135; P.254; 136; P. 620, 137; P. 146, 138; 113, 139; P. 25, 140; P. 93-94], much less often of the 12 series with Keggin structure [141; P. 56, 142; P. 58, 143; P. 92, 144; . 163]. Many works are known, where other salts of the molybdocobalate type [145; P.548-549] containing mainly heteropolyanion $[Co_2Mo_{10}O_{38}H_4]^{6-}$ [146; P. 41-42, 147; P. 45-47, 148; P. 258] are taken as a basis. The influence of the Co(Ni)/Mo ratio in catalysts obtained through the heteropolyanion formation stage $[Co_2Mo_{10}O_{38}H_4]^{6-}$ using different chelating agents was investigated [149; P.69, 150; P.28].

A new direction in the synthesis of hydrotreating catalysts based on the preliminary preparation of individual bimetallic chelate complexes with their subsequent deposition on traditional carriers has been developed [151; P. 5; 152; P. 509, 153; P. 279]. The specificity of the synthesis of such catalysts is the selection of the heat treatment regime, providing the preservation of complex compounds, up to the stage of sulfidation of the oxide form. In this case, sulfidation is preferably carried out with a mixture of hydrogen and hydrogen sulfide. The most popular chelating agent for this type of catalyst is citric acid, due to its commercial availability and pronounced positive effect on the hydrodesulfurizing activity 154; C.109; 155; C. 25-27, 156; C. 885-887, 157; C. 175-176]. The possibility of joint use of heteropolyacids and citric acid has been revealed [158; P.109, P. 112-114], according to Raman spectroscopy (Raman spectroscopy) the combination of organic and inorganic acid in the impregnation solution reduces the degree of depolymerization of heteropolyanions when adsorbed onto the carrier surface.

A number of publications [117; P.446, 118; P.915, 160; P.55] are devoted to comparing the influence of impregnation methods on the uniformity of distribution of hydrogenating metals over the volume of granules and catalytic properties of catalysts. It is shown that the ANM hydrotreating catalyst obtained by sequential introduction of active components is superior in activity to the catalysts obtained by coprecipitation of active components and aluminum oxide hydrate. It is noted that the deposition of molybdenum on the carrier is a complex process that consists of at least two stages: 1-dissociation (or association) of ammonium paramolybdate upon dissolution into different structurally different ions ($MoO^{2-}{}_4$, $Mo_4O_{(13)}({}^{2-})$, $Mo_7O_{(24)}({}^{2-})$, etc.) depending on the salt concentration and pH of the medium. 2-adsorption of molybdate structures with different degrees of association on the carrier.

It has been experimentally proved that the molybdenum content in the intermediates increases with increasing impregnation time and concentration of impregnation solution, and the degree of adsorption of molybdenum salt on the carrier tends to a certain limit. This is caused by the increase in the size of molybdenum polyanions in solution with increasing salt concentration in solution and decreasing pH. Therefore, molybdenum in the catalyst is distributed unevenly, most of it is located in the surface layer of the carrier. The increase in the average pore radius is explained by the initial filling of narrow pores and, consequently, by the corresponding increase in the proportion of large pores

One of the most important factors affecting the characteristics of ANM catalysts is the method of nickel and molybdenum deposition on the carrier, in particular, the order of impregnation of $Al_3O_3$ granules with salts of these elements. By the methods of CR spectroscopy, electron diffuse reflectance spectroscopy (EDSR) and X-ray photoelectron spectroscopy (XPS) it is shown that during the impregnation of the carrier with a mixture of salts a relatively larger amount of $Ni^{(2+)}$ ions is formed than in two-stage variants of deposition, the ligand sphere of which includes $O^{2-}$ ions bound to aluminum ions. If nickel is applied to a surface already coated with molybdenum compounds, the bulk of $Ni^{(2+})$ ions interacts with $Mo_4^{2-}$ anions already through the oxygen ligand to form A1-O-Mo-Ni-O-Mo-O-O-A1 structures

The study of the nature of hydrogenating compounds by spectral methods showed that when impregnating the carrier with hydrogenating metal salts, mainly surface poly compounds of molybdenum with inclusion of nickel are formed, which weakly interact with the carrier. The interaction of molybdenum and nickel with the carrier increases with increasing calcination temperature

In industry quite often used, technologically attractive, is the method of mixing wet aluminum hydroxide with solutions or solid salts of all hydrogenating metals. This method, with various variations, has been discussed in some detail in papers and patents [110; 1265-1266, 161; P. 1513-1514]. A mixture of solid molybdenum and nickel (or cobalt) containing salts is introduced into aluminum hydroxide, using in some cases nitric or phosphoric acids as peptizers. Sometimes a wet cake of freshly precipitated aluminum hydroxide is also mixed with a hot solution of nickel and molybdenum salts. It was found that $MoO_3$ in an amount of up to 8.6%, introduced in the form of ammonium paramolybdate, increases the surface area of the catalyst compared to $\gamma$-$Al_2O_3$, while NiO, introduced in the form of nickel nitrate, on the contrary, slightly reduces it

The influence of the type of carrier raw on the degree of interaction of active components with each other was also investigated: samples obtained by mixing bimetallic solutions with bemitite [111; P.178-179], a mixture of pseudobemite and gibbsite [110; 1264-1265, 112; P.190-193], bayerite and even $\gamma$- $Al_2O_3$ [162; P183-185] were compared.

The main problem of synthesis of effective ANM and ACM catalysts for hydrogenation is the selection of such a ratio between the content of these compounds at a fixed chemical composition, which provides the optimal activity of sulfidation products of catalysts in hydrotreating reactions

Such information can be obtained by selective removal of individual compounds or groups of compounds of active components from ANM catalysts and subsequent comparison of catalytic properties of initial catalysts and separation products of structural states of Ni-Mo components. It should be noted that the extraction process was carried out in different ways [162; P.178-180., 111; P.178-181; 112; P.187]. Molybdenum was extracted from the catalyst composition by daily soaking 0.3 g of the sample in 30 ml of water [114; P.456-457.].

In [162; P.178-180], a 2 g catalyst suspension was poured into 10 ml of water and kept at room temperature for 4 h; then the solution was drained and the catalyst was calcined. Then the similar procedure was repeated 2 more times with the same catalyst suspension and new portions of solvent. The resulting 30 mL of solution was evaporated, the dry extract was weighed and subjected to further study, and the catalyst was calcined at 823 K for 4 h, then the extraction was repeated. Four-fold extraction of the same catalyst suspension with intermediate calcination allowed to extract practically all water-soluble material. Multiple treatment of calcined ANM catalysts with water with intermediate calcination in air at 550° C allows to extract of Ni contained in them and 10-50% of $MoO_3$, which after extraction and crystallization is a mixture of $NiMoO_4$, $Al_2(MoO_4)_3$ and $MoO_{(3)}$. The amount of Mo removed increases dramatically with aqueous ammonia extraction compared to aqueous extraction, whereas the amount of Ni removed is essentially unchanged. After preliminary aqueous extraction, additional treatment of the catalysts with 1 N ammonia solution results in the removal of molybdenum only.

The issues of optimization of porous structure in the synthesis of catalysts, along with the distribution of active components, are of paramount importance, so they have been given much attention in the literature for many years [163; P.931-933, 164; 1399-1405,165; P.602-604, 166; P.53-54]. The aging conditions (time, pH, temperature) and the type of intermicellar liquid, also determine the nature of compaction (shrinkage) of gels, and hence their porous structure. Substitution of interlayer water with methanol before drying leads to loss of surface tension, which prevents skeletal compression and, as a consequence, preservation of the size of crystals and, consequently, of the pores between them. On the other hand, the addition of burnout pore-forming additives depending on their type (ketones, esters, water-soluble polymers, starch, etc.) and their amount allows to regulate the pore volume (e.g., from 0.3 to 1.0 $cm^3/g$) and pore size distribution [167; P.44]. This is especially important in the development of catalysts for the processing of heavy distillate and residual feedstocks. At high pressures, which are required to achieve a deep degree of hydrotreating, the heavy feedstock is partially condensed, so that the hydrotreating process takes place in the catalyst grain partially filled with the liquid phase. In the work of Smolikov et al. on the example of $MoS_2/\gamma\text{-}Al_2O_3$ catalyst, it was shown that the character of microdistribution of active component inside the mesopores is approximately the same, i.e. the sulfide phase is distributed over the whole range of mesopores present in the carrier. The correspondence between macro- and micro-distribution is due to the fact that the conditions of the impregnation stage, which promote the uniform spreading of the sorption front over the carrier granule, are simultaneously also the conditions that ensure diffusion and anchoring of the active component in the entire range of pores and, thus, provide the equalization of the heterogeneous micro-distribution of the active center in the catalyst [168; P.54-55]. Due to the need in wide-porous materials for the synthesis of modern catalysts, special technologies for the synthesis of mesoporous carriers have been developed: pseudobemite [169; P.276], aluminum oxide [170; P.70-71, 171; 550-553], and aluminosilicates of the Al-SBA-16 type [172; P. 1-4]. On traditional alumina-oxide carriers, it has been proved that with increasing the volume of large pores from 0.1 to 0.5 $cm^3/g$, the degree of utilization of whole grains in the hydrogenation of impure asphaltenes increases from 0.16 to 0.5. However, the reaction rate of desulfurization of unassociated sulfur compounds is practically unchanged. The role of active components in the genesis of porous structure can be clearly seen on the example of the deposition of molybdenum and nickel compounds. Namely, the decrease in the specific pore volume and specific surface area of ANM samples, compared to the original carrier. Blocking of the smallest pores by active metal compounds is confirmed by electron microscopy data.

The article by Lurie M.A. and co-authors [164; P.1400-1402] is devoted to the issues of hydrotreating of heavy oil feedstock on catalysts with different porous structure. Carriers for catalysts were prepared using crushed aluminum oxide or burnout additives. The ANM catalyst on the optimal carrier had the following pore volume distribution by radii: 3-10 nm (0.21 $cm^3/g$); 10-100 nm (0.1 $cm(^3)/g$) more than 100 nm (0.08 $cm^3/g$). Hence, it was concluded that there is no correlation of hydrodesulfurization activity with the number of pores with radius less than 10 nm and proved the increased stability of large-porous catalysts in the process of desulfurization of heavy oil feedstock, due to more efficient use of the active component and reaction volume.

It is accepted that the texture of carriers (presence of a developed system of large pores from 6 to 10 nm for hydrodessulfurization of diesel fractions and more than 10 nm for vacuum gas oils and residual oils, on the one hand, determines the availability of active catalyst centers for large molecules, substituted alkyl dibenzothiophenes and resinous asphaltene substances, the geometric sizes of which reach from 70 to 100Å (from 7 to 10 nm) [66; P.79, 160; P.105].

On the other hand, uniform deposition of high concentrations of transition metals of group VIII and VIB, of the order of 6 molybdenum atoms per $nm^2$, requires a large specific surface area (from 200 to 350 $m^2/g$) of the carrier and, consequently, a sufficiently developed system of small pores. A possible compromise for two opposing opinions on the porous structure issue was the research and synthesis of carriers with bimodal distribution of pores in a rather narrow range of radii. Chukin G.D. singles out three intervals of pore radii for research: 1-2nm (micropores), 2-5nm (transition pores) and 5-10nm (mesopores) [84; 100]. It has been experimentally proved that pores up to 2 nm are microcracks at the edges of primary aluminum oxide crystals, and pores over 5 nm represent the space between primary crystals (secondary pores). The pores of these groups have fundamentally different nature and mechanism of change during heat treatment, so for the synthesis of carriers with a given texture, the issues of synthesis of initial aluminum hydroxide and calcination conditions for the phase transition of the precursor hydroxide into aluminum oxide are of crucial importance [84; P.109, 160; P.28-87]. For example, [173; P.1765-1767], a sample of aluminum oxide carrier with bimodal pore size distribution (Deff.1 = 50Å and Deff.2 = 125Å) was prepared. The authors showed higher catalytic activity of the sample with bimodal distribution of pore sizes compared to the industrial sample, but the choice of the values of effective diameters is not substantiated either theoretically or experimentally. The most common carriers having sufficiently high for modern industrial reactors mechanical strength and necessary texture are: aluminum oxide [174; .1], aluminum oxide modified by additives of natural or synthetic aluminosilicates [175; .1, 176; C.1, 177; C.1], titanium dioxide [178; C.1], boron [179; C.1, 180; .1] and other modifiers. Of particular interest are carbon and carbon-containing carriers [160; P.50-89, 181; P.858], which make it possible to minimize the interaction of active components with the carrier and maintain a given ratio of Co/Mo or Ni/Mo in the sulfidized form.

From the analysis of literature sources it follows that one of the most accessible methods of obtaining durable catalysts with bidisperse porous structure is the combination of fine powder of aluminum hydroxide with coarse powder of aluminum oxide or crumb of ACM catalyst in the molding mass [182; P.1, 183; P.1,184; P.1]. The catalysts intended for processing of high-sulfur heavy residual feedstock are preferably modified with boron and phosphorus compounds [185; P.47-48, 62, 186; P.321, 187; P.153-155]

## §1.5 Reactivity of sulfidized catalysts in the conversion of typical sulfur compounds of oils

The choice of catalyst composition is determined by many factors that positively characterize them both from the standpoint of their electronic structure and the properties of their salts and compounds, which determine both the manufacturability of catalyst creation operations and the applicability in practice of the created catalytic system. A special place is occupied by studies of the sulfidized form of catalysts and the sulfurization stage of the oxide form of active metals. About forty years in the literature different models of active phase are compared and their correspondence to experimental facts is discussed [188; P.43, 189; P.395-398, 190; P.201-202]. One of the first models of active phase - "monolayer" assumed the presence of promoter in the volume of the carrier, according to this model, the maximum activity is possessed by surface molybdenum sulfides in different coordination, but the activity of promoter compounds of spinel type is low. With the development of physicochemical methods for the study of catalysts, including in-situ [188; P.43], it became possible to study the extremely complex chemistry of sulfide catalysts and, as a consequence, various models of the active phase were considered. The pseudo-intercalation (introduction) model involves decorating the sulfide phase with promoter atoms and modifying additives [191; P.89-92, 192; P.217,221-224], contact synergism or remote control model, rib-rim model and others [188; P.43]. A group of researchers led by H.Topsoe since 80s [189; P.412-420, 190; P.199] was engaged in the study of morphology and the role of Co-Mo-S type structures in the activity of sulfide catalysts [193; P.3-8, 194;34-36, 195; P.86-88, 196; P.195, P.202-203]. It has been found that the active component of Co(Ni)Mo(S)/$Al_2O_3$ catalysts are small $MoS_2$ crystallites, which are short layered packing. According to the researchers, the most active catalytic centers are Co(Ni) atoms bound by sulfide bridges to the surface of the layered packing of $MoS_2$ crystallites, which they designated as the "Co-Mo-S phase". According to this theory, the active centers are located on the surface of the edges, contain coordination-unsaturated sulfur atoms and Mo atoms. Under reaction conditions such structures are not stable, hydrogen reacts with sulfur to form coordination-unsaturated centers or anionic vacancies. That is, the active centers are sulfide vacancies. However, it follows from the results of [197; P.519-521] that the active centers, in addition to sulfide vacancies, can also be fully coordinated with sulfur centers, the properties of which are similar to metallic ones. Although it was initially believed that depending on the type of interaction with the carrier, the Co-Mo-S phase, where the promoter sulfide decorates the edges of molybdenum disulfide, can be realized as a type I or type II structure with different catalytic properties, currently isolated more fully sulfidized phase of type III. N.Startsev and co-authors proposed a model of sulfide bimetallic compound, which is a variant of the "Co-Mo-S" model as applied to a carbon carrier (sibunite) [160; C38-54]. In the review article by Kogan V.M. and co-authors [198; P.638-639] an assumption is made that the energy of metal-sulfur bonding on the surface of the sulfide cluster depends on the coordination number of surface metal atoms on sulfur and the position of the metal in the Periodic Table. And sulfur mobility in the sulfidized catalyst, according to the authors, is a key parameter determining the activity of catalysts in hydrodesulfurization and hydrogenation. The study of the influence of metal nature on the catalytic properties of catalysts (on the same carrier and equally sulfidized) in the thiophene conversion reaction allowed the authors to conclude that the productivity of active centers increases in the series Co< (Co+ Ni) <Ni. Radioisotope method proved that SH-groups of Ni-sulfide particles show higher reactivity in the formation of $H_2S$ in the hydrogenolysis of thiophene than similar SH-groups, but associated with cobalt [198; P.655]. From the above mentioned, the importance of the technological stage of sulfidation of hydrotreating catalysts follows, the main task of which is the conversion of active phase precursors

into the sulfide state. There are different variants of the sulfidation process, but two are fundamentally different - liquid-phase and gas-phase. For liquid-phase method light oil fractions with increased sulfur content (due to the introduced sulfidizing agent) are used, and for gas-phase sulfidation a mixture of hydrogen sulfide with some  is used. Tuxen A. et al [199; P.349-351], with the help of a complex of physicochemical methods, convincingly showed that at liquid-phase sulfidation there is an introduction of carbon into the structure of the active phase of molybdenum disulfide. This leads to a decrease in the activity of the catalyst and reduces the stability of the sulfide phase, hence, gas-phase sulfidation, especially in an inert gas environment, is preferable for obtaining more active catalysts in the sulfide form [200; P. 77-79].

According to the generally accepted opinion [188; P.42], aliphatic sulfur compounds easily enter into hydrogenation reactions in the presence of hydrogen and a catalyst, while the C - S bond is homolytically broken and the free valences are saturated with hydrogen. All sulfur compounds of oil are conditionally divided into easily hydrogenated mixture of aliphatic and cyclic sulfides, medium hydrogenated - homologues of thiophene and benzthiophene, and, finally, to difficult hydrogenated include derivatives of dibenzothiophene. It was found that in the composition of aromatic sulfur compounds in fractions boiling above 300°C there are $C_2$-$C_5$ - substituted benzothiophenes and dibenzothiophenes, dibenzothiophene, 4-methyldibenzthiophene and dimethyldibenzthiophenes. In the literature review by Y. V. Yeremina. [131; P.76-89], the results of studies showing that the introduction of a methyl substituent in dibenzthiophene decreases the degree of hydrogenation by 30%, and the introduction of two methyl substituents reduces the degree of conversion by 80%. It follows that the main factor determining the depth of hydrodesulfurization is not the total content of sulfur in the raw material, but the concentration of difficult to hydrogenate compounds. The reactivity of certain groups of organosulfur compounds decreases in the following sequence, mercaptans> disulfides> sulfides≈ thiophenes> thiophenes> benzothiophenes> dibenzothiophenes [188; P.42] sulfur-containing compounds to hydrogenolysis with increasing number of aromatic and naphthenic rings in the molecule increases. If the methyl groups are not located in close proximity to the sulfur atom, the reactivity of such an organosulfur compound increases relative to an organosulfur compound containing no substituent. When the substituent is located close to the sulfur atom, the reactivity decreases due to steric effects. Although knowledge of the group composition of sulfur-containing compounds of the processed feedstock is necessary for proper selection of the catalyst, so far the establishment of the relationship between the composition of organosulfur compounds and catalytic activity is empirical in nature. Compounds of thiophene series are among the most stable organosulfur compounds of oil, therefore thiophene is often chosen as a model substance for studies of catalyst activity. A great deal of scientific work has been devoted to the study of the mechanism of thiophene hydrogenolysis [201; . 419, 202; P. 116-118, 203; P. 94-96, 204; P 34, 205; P. 61-62]. Three routes of thiophene hydrogenolysis reaction were discussed in the literature: 1) detachment of sulfur atom with formation of hydrogen sulfide and butadiene; 2) complete hydrogenation of thiophene ring with subsequent breaking of C-S bonds; 3) partial hydrogenation of thiophene, detachment of sulfur atom as hydrogen sulfide, formation of butenes. The final products of the interaction are saturated hydrocarbon and $H_2S$. The formation of tetrahydrothiophene was found as an intermediate product by ICS method by a number of authors when $MoS_3$ was used as a catalyst. The possibility of breaking C-S bonds in thiophene in the presence of hydrogen without saturation of two double bonds was also considered, the proof of realization of this variant is the presence of butadiene among the conversion products of thiophene. The paper [188; P.46] emphasizes the contribution of Torsoe's work in the study of reaction routes (hydrogenation and direct hydrodesulfination). In the products of hydrogenation of benzthiophenes at elevated hydrogen pressure, besides ethylbenzene, dihydrobenzothiophene was found, so it was suggested that this reaction proceeds sequentially through the formation of dihydrobenzothiophene and ethylbenzene. Experimentally [188; P.46] an increase in bond strength in the series 46-dimethyldibenzothiophene <4-methyldibenzothiophene <dibenzothiophene <2,8-dimethyldibenzothiophene has been proved [188; P.46]. Some researchers attributed the lower rate of desulfurization of 4,6-dimethyldibenzothiophene to the poor adsorption capacity of the catalyst surface, while others attribute it to steric hindrance on the part of methyl groups. From the analysis of numerous works devoted to the establishment of the mechanism of reactions of hydrodesulfurization of thiophene and its derivatives as the most important heteroatomic components of oil feedstock, the conclusion about the importance of stages and routes of hydrogenation was made. Moreover, as the molecules become more complex, i.e., in the process of transition from thiophene to benzthiophene, its substitutes and, especially, 4,6-dialkyldibenzthiophene, the hydrogenation stage in the route of "direct hydrodesulfurization" becomes more and more important [188; P. 50]. According to the results of the analysis of the experience of catalysts operation on different types of raw materials, dibenzthiophene transformations on ACM catalysts proceed mainly by a fast reaction with C-S bond breaking. The hydrogenolysis of dibenzthiophene through the hydrogenation stages of the aromatic ring (slow reactions), along with easy hydrodesulfurization of thiophene and its homologues, preferably proceeds on ANM catalysts [206; P.298, 207; P.170,172, 208; P.1366-1367, 209; P.418-419]. It has been proved that in the case of spatially hindered 4,6-dimethyldibenzothiophenes and 4,6-dimethylbenzothiophenes, when methyl groups physically obscure the sulfur atom from contact with active centers on the catalyst surface, the reaction with C - S bond breaking proceeds at a very low rate. Therefore, the ACM catalyst is not efficient. However, hydrogenation of the aromatic ring, which allows one of the methyl groups to move away from the sulfur atom, followed by C - S bond breaking proceeds much faster. Since this sequence of reactions is better catalyzed by ANM catalyst, its application is more efficient than that of ACM, provided that the partial pressure of hydrogen required to realize the hydrogenation step is adequate. Thus, considering that the calculated hydrogen consumption for the reaction of dibenzthiophene hydrogenolysis, depending on the sequence of stages varies by 3-5 times, the correct selection of the catalytic system can regulate the need for hydrogen.

## §1.6 Evaluation of catalyst activity

Testing of catalyst activity on a high-pressure flowing unit. The test was carried out using raw materials - III-fraction of oil distillate and deasphaltized residue. Drying of the catalyst is carried out in the reactor. For this purpose the reactor is heated in a hydrogen current up to 200° C with a temperature rise rate of 50° C/h, at a pressure of 0.5 MPa for the first 10 hours, then raised to 1.5 MPa. Sulfidation of the catalyst is carried out by raw materials according to a special schedule. At the end of sulfidation at 280° C, at a volumetric rate of feedstock 2 h-1 raise the temperature to 330° C and the pressure to 3.0 MPa. Simultaneously with temperature increase, hydrogen circulation is increased up to 300 $m^3/m^{(3)of}$ feedstock. After sulfurization, the volumetric feed rate of the feedstock is reduced to 1.0h-1. To evaluate the effectiveness of the tested contacts on the parameters of the oil hydrotreating process, the following main characteristics were determined: color, kinematic viscosity, viscosity index, sulfur content, density, refractive index, etc. according to GOSTs in force at the Fergana Refinery.

Catalyst activity tests carried out under the following conditions: Operating pressure, MPa - 3.0; Temperature,° C - 350; Feedstock feed rate, h-1 - 1.0; Ratio H: feedstock, $m^3/m^3$ - 300; Duration of the experiment, hour - 720; Sampling frequency of hydrogenizate, hour - 4.0.

The scheme of the unit is presented in Figure 1.2. The III-fraction oil from the feedstock tank 4 is fed by pump 5 into the reactor 6, hydrogen from the cylinder 1, the pressure of which is maintained by the gas flow meter 2. Gas-liquid feedstock stream passes through a layer of heated catalyst, where the process of its hydrotreating takes place. The gas-liquid stream of the hydrogenizate then passes through the cooler 12 and enters the separator 16, where gas and liquid are separated. The separated gas passes through a gas meter 15 and is then neutralized with an alkali solution. The hydrotreated product enters the receiving vessel 16.

Selected samples of hydrogenizate are washed with 10% sodium hydroxide solution to remove hydrogen sulfide and from 3 samples make one average for sulfur content analysis.

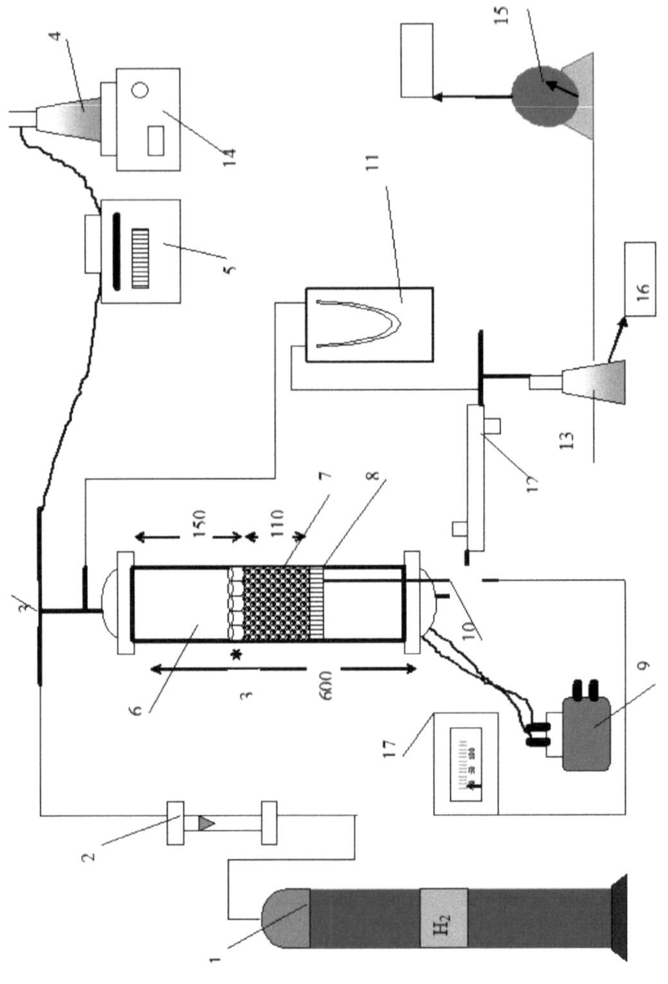


*1-hydrogen cylinder, 2-rotometer, 3-mixer, 4-capacity for feedstock, 5-peristaltic pump, 6-reactor, 7-test samples, 8-lattice, 9-latre, 10-thermocouple, 11-manometer, 12-cooler, 13-separator, 14-magnetic stirrer, 15-gas meter, 16-receiving vessel, 17-millivoltmeter*

**Fig. 1.2. Technological scheme of the pilot plant**

According to the results of the analysis determine the degree of desulfurization of raw materials and find the activity of the catalyst (in %) according to the formula:

$$X = \frac{(V - V_1)X \bullet 0,0008 \bullet 100}{G},$$

where: V-volume of exactly 0.05 normal hydrochloric acid solution used for titration in the control experiment, ml; V1-the same in the target experiment, ml; 0.0008-quantity of sulfur equivalent to 1 ml of exactly 0.05 normal hydrochloric acid solution, g; G-suspension of the test product, g; - X-content of total sulfur in the analyzed product, %;

Catalytic activity is calculated by the formula:

A = 100-X, where A - catalytic activity, %

Determination of the content of sulfur compounds in the hydrogenizate was determined by the accelerated method (GOST 1437-75). The essence of the method consists in combustion of a sample of petroleum product in a stream of air, capture of the resulting sulfur dioxide and sulfur dioxide with a solution of hydrogen peroxide with sulfuric acid and titration with caustic soda.

## §1.7 Quality assessment and preparation of raw materials

It is known that the most important parameters of carriers and catalysts directly depend on the properties of the initial components, especially in the case of natural minerals characterized by heterogeneity of composition. Therefore, increased attention is paid to the evaluation of the quality of raw materials - a necessary condition for obtaining carriers with specified physical and chemical characteristics. We studied two grades of kaolin. Angren gray unenriched kaolin (SK) is characterized by small particle sizes of clay mineral, so the molding masses are characterized by good plasticity (Fig.1.3). Its disadvantage is the presence of impurities, primarily iron and large quartz crystals. Enriched white kaolin (AKF-78) contains more aluminum oxide, an order of magnitude less harmful impurities, but consists of larger particles, which worsens the conditions of extrusion of the molding mass and adversely affects the mechanical strength. The conformity of kaolin to the requirements for use as a catalyst component is determined by wet and dry sieving, fractions larger than 0.2 mm are not reasonable to use. Waste fiber production "nitron", contain in their composition the following functional groups: $-NH_2$, -COOH, -COONa, -CN, -ON, $-COOSH_3$. To modify the carrier it is allowed to use only raw materials with minimum content of sodium ions, mineral residue of which should not exceed 0.1 wt. %. Polyvinyl acetate is used in the form of an aqueous dispersion containing no mineral additives.

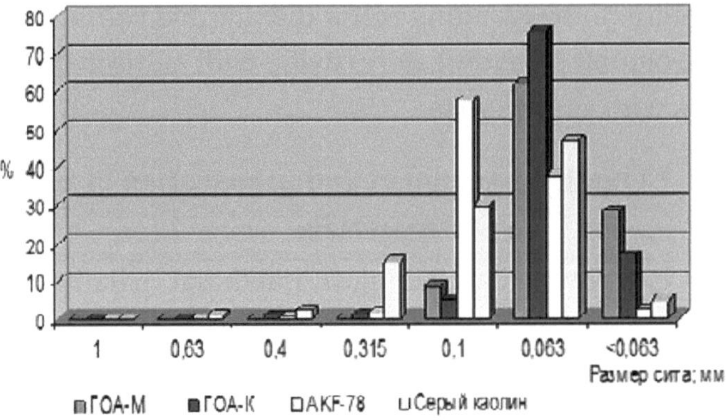


**Fig.1.3 Results of sieve analysis of aluminum hydroxide and kaolin samples**

The largest component by mass is aluminum hydroxide, which is not produced in Uzbekistan [211; P. 122], so for catalyst synthesis it is necessary to use only imported product in the form of powder or "cakes". Taking into account economic indicators, the ability to flexibly adjust the texture and maximum reactivity of freshly precipitated hydroxides, developers and manufacturers of catalysts seek to organize the technological chain, bypassing the stage of drying of aluminum hydroxide precipitate.

The methods of preparation of new generation catalysts are focused on freshly precipitated hydroxide with a certain morphology, and with great difficulty are reproduced on dried aluminum hydroxides supplied by various manufacturers. The texture and phase composition of aluminum hydroxides is highly labile and depends on the conditions of transportation and duration of storage, therefore the incoming inspection of imported or illiquid raw materials is of paramount importance. Pseudobemitic (P/be) modification is the most susceptible to environmental influences. For the purpose of selection of raw materials, samples of aluminum hydroxides were taken from potential suppliers from warehouses in Tashkent and their consumer properties were evaluated [212; P.28-30].

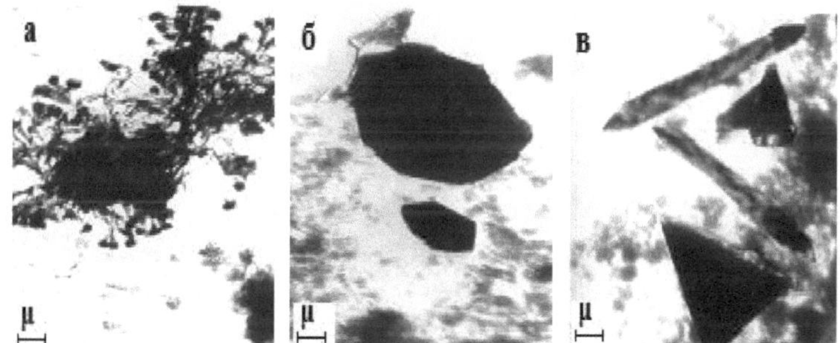


**Fig. 1.4 Electron micrographs of fragments of aluminum hydroxide samples: a - fresh pseudobemite; b - sample № 4 c - sample No.6**

A sample of fresh pseudobemite powder (sample No.1 in Table 1.1) was chosen as a reference. This sample is characterized by: homogeneity of phase composition, high dispersibility and good mechanical strength of granules, however, the quantity of this sample does not exceed several kilograms. In the composition of fresh aluminum hydroxide powder according to derivatography and ICS data there are adsorbed ions NH4+ and NO3-, stabilizing the pseudo-beamite modification. It was found that under the influence of high summer temperatures, stabilizing ions and interlayer water are easily removed from the composition of aluminum hydroxide.

The acidic character of the aqueous suspension of the fresh powder draws attention, the pH value of which serves as a first test for the suitability of this raw for extrusion molding.

**Table 1.1.**

**Characteristics of investigated samples of aluminum hydroxides**

| **Parameters** | **Aluminum hydroxide sample numbers** | | | | | |
|---|---|---|---|---|---|---|
| | **№1** | **№2** | **№3** | **№4** | **№5** | **№6** |
| Phase composition (crystal size; nm) | P/be (1.5 - 3.0) | P/be (8.0) | P/be+ BE (10.0-25.0) and traces of BA | BE(20.0) BA(25.0-100.0) | BA(25.0) G(28.0 - 300.0) BE(20.0-100.0) | Г > 100.0 |
| pH of aqueous suspension 1:50 | 4.1 | 4.3 | 4.4 | 5.7 | 5.9 | 7.5 |

| Parameters | Aluminum hydroxide sample numbers | | | | | |
|---|---|---|---|---|---|---|
| | **№1** | **№2** | **№3** | **№4** | **№5** | **№6** |
| Specific surface area; $m^2/g$ | 102.8 | 89.5 | 85.3 | 80.3 | 77.5 | 25.2 |
| Losses on ignition; % | 40.1 | 37.8 | 32.6 | 27.3 | 6.49 | 7.76 |
| $NO_3^-$ content; % | 0.37 | 0.15 | 0.05 | 0 | 0 | 0 |
| Solubility in 5N H $NO_3$;% | 70-73 | 68-72 | 60-65 | 45-56 | 22-38 | 10-12 |
| Aluminum oxide granules with a diameter of 5 mm (peptization at Mk= $5\cdot10^{-4}$) | | | | | | |
| Mechanical strength; kG/gran. | 4.91 | 4.85 | 3.6 | 2.6 | 2.1 | Not moldable |
| % of rejected pellets | < 1 | 2-3 | 5-8 | 10-13 | 20-30 | |

Fresh powder of sample 1 of the investigated aluminum hydroxide is a poorly occrystallized pseudobemite with primary crystal sizes of 25-30Å (Table 1.1.1), stabilized by nitrate ions, which cause a decrease in pH when the sample is suspended in water. In the electron microscope, the morphology of samples 1 and 2 can be seen as aggregates of dense particles with a diameter of about 60Å and needle-like formations with a length of 60Å; the admixture of amorphous phase (particles less than 30Å) appears as loose aggregates with a size of several hundred angstroms.

At lower magnification, larger dense irregularly shaped fragments and open-work cluster-like particles formed by the mechanism of oriented accretion of primary microparticles are observed (Fig. 1.4 a). These characteristic micro-particle aggregations are responsible for the formation of the primary porous structure of active aluminum oxide  Typical pore distribution of aluminum oxides based on samples #1-2: mesopores with radii from 37 to 93Å (maximum around 50Å) occupy a volume of 0.45 $cm^3/g$, there are also macropores with radii 500-1500 and 1800-5000Å, with volumes of 0.02 and 0.08 $cm^3/g$, respectively. The study of surface properties revealed a decrease in the concentration of strong ($pK_a \leq -5.6$), weaker acid centers ($pK_{(a)} \leq -3$) and basic centers with $pK_{(a) \geq +9}$ on sample 2, compared to #1. During the development of an alumina-kaolinborate carrier based on a similar pseudobemite sample for the ANM 2/5 catalyst [213; P.55,60.], it was proved necessary to increase the acid modulus ($M_{k)}$ from $5\text{-}10^{(-4})$ (optimal for peptization of fresh hydroxide) to 2÷ $3\text{-}10^{-3}$ g-mol $HNO_3$/g-mol $Al_2O_{(3)}$. With time, the powder aging processes deepen, similar to the aging of precipitates in the mother liquor, although strongly retarded. The first negative moment, which technologists face when using large-crystalline modifications of aluminum hydroxide, is an obvious decrease in plasticity of the mass, forming and cutting of granules is difficult due to the separation of liquid when the mass passes through the extruder (Table 1.1). To obtain quality pellets from samples № 2 and 3 it is necessary to increase the acid modulus.

Fluctuations in phase composition occur even within one batch of raw materials. Thus, in the XRD of the sample taken from the bag near the warehouse wall, clear reflections typical for bayerite (BA) and gibbsite (G) are detected, while the phase composition of the powder sample from the other end of the same package is represented mainly by the bemitic modification (BE), with different degrees of oxidation. When viewing sample 3 in the electron microscope, the appearance of large aggregates in the form of plates of 700-1000 Å, characteristic of bemitite (Fig. 1.4 b) and a decrease in the proportion of cluster-like openwork formations of pseudobemite (Fig. 1.4a) were recorded. The diameter of primary particles increased to 70-130 Å. In addition, the appearance of individual triangular particles with sizes of $10^4 \div 10^5$ Å, characteristic of coarse-crystalline bayerite, was noted. Large formations prevail in samples No. 4-5 (Fig. 2.3 b), pseudobemite is not found in their composition, and sample No. 6 is represented by large-crystalline gibbsite (Fig. 1.4 c). All these samples are not amenable to extrusion molding at the previously found optimal acid modulus.

With the purpose of express estimation of reactivity of aluminum hydroxides, all samples were treated with nitric acid solutions, varying the acid modulus from 0 to 5 g-mol $HNO_3$/g-mol $Al_{(2)}O_3$ at the ratio of liquid to solid phase 50:1 and their solubility was determined. It was found that samples having morphology intermediate between pseudobemite and crystalline bemite are able to disaggregate in nitric acid solutions [212; P. 27]. Improvement of formability of the mass prepared from powders No. 4 and No. 5, predominantly bemitic modification, with high pH of suspension is achieved when the acid modulus increases above $2 \cdot 10^{-2}$ g-mol $HNO_3$ / g-mol $Al_2O_3$. In addition, the carrier is dominated by pores with a radius less than 30 Å, which are generally ineffective in hydrotreating processes. Such batches of aluminum hydroxide (№ 4-5) can be used only as an additive (not more than 20%) with careful mixing with powder having low pH suspension, and peptization should be carried out at acid modulus $4 \div 8 \; 10^{-3}$.

Considering that acid treatment, even with a modulus of about five, has no appreciable effect on the size and shape of crystals of large-crystalline bayerite and gibbsite, it is reasonable to reject aluminum hydroxide powders with a suspension pH higher than 5.5 and solubility less than 30%, with an acid modulus of 1.4 g-mol $HNO_3$ / g-mol $Al_2O_3$

Thus, the criteria for assessing the quality of non-liquid aluminum hydroxide for rational use as a component of catalyst for hydrotreating of oil fractions have been developed

# CHAPTER II. CARRIER SYNTHESIS TECHNOLOGIES

## §2.1 Formation and study of carriers with bidisperse porous structure

As follows from the literature review, catalysts for hydrotreating of vacuum distillates and deasphalted oil residues require carriers combining mechanical strength and developed system of large pores (sizes of about 10÷40 nm) necessary for operation in modern reactors. Therefore, the main purpose of this part of the research is to develop a carrier with a developed system of wide pores for synthesis of catalysts for hydrotreating of highly viscous oil distillates, necessary, first of all, to increase the service life. Samples No. 3 and No. 5 (Table 1.1), purchased in sufficient quantity and designated hereinafter as GOA-M and GOA-K, respectively, were selected for research and development of carrier preparation.

In the process of preparation of the samples presented in Table 2.1, it was revealed that nitric acid is the optimal peptizer for preparation of mechanically strong carrier granules, while ammonia is less effective. Carrier granules containing GOA-M (with particle size from 0.004 to 0.04 mm) and gray kaolin (with particle size from 0.004 to 0.04 mm) are superior in their mechanical properties to the samples obtained with white enriched kaolin AKF-78 (with particle size from 0.004 to 0.04 mm), but are significantly inferior in porosity. On the basis of coarse-dispersed GOA-K (particle size from 0.04 to 0.1 mm) by this method it is not possible to achieve satisfactory results. The introduction of boric acid into the molding mass is accompanied by an increase in strength and a decrease in the proportion of large pores. Taking into account the wide application of zeolites for the synthesis of hydroprocess catalysts, we tried to introduce some crystalline aluminosilicates into the developed carrier.

**Table 2.1.**

**Influence of synthesis conditions on physicochemical characteristics of carriers**

| № | Media components | | Mechanical Strength, | | Summarized pore volume; $cm^3/g$ |
|---|---|---|---|---|---|
| | Aluminum hydroxide, % | Silicon-containing additive, ( %) | Crushing on the end, MPa | Scrubbability in 15 min, % | |
| 1 | GOA-M | 0 | 1,5 | 6,38 | 0,65 |
| 2 | GOA-C | 0 | Not moldable | | 0,60 |
| 3 | GOA-M | AKF-78 (20) | 1,4 | 4,85 | 0,54 |
| 4 | GOA-M | Gray kaolin (10) | 1,7 | 6,54 | 0,50 |
| 5 | VOA-K | Gray kaolin (20) | 1,8 | 6,35 | 0,43 |
| 6 | GOA-M | AKF-78 (10) | 1,6 | 5,51 | 0,56 |
| 7 | GOA-M | Zeolite NaX | 0,8 | 12,8 | 0,60 |
| 8 | GOA-M | Zeolitized clay | 1,2 | | 0,64 |

The introduction of crushed NaX powder of sample No.7 into the composition of the carrier reduces the abrasion resistance of the carriers. The gel-like mass formed under controlled conditions of zeolitized clay synthesis, on the contrary, hardens granules and increases the total pore volume [214; P.83, 215; P.226]. Based on the results of adsorption of large test molecules [210; P. 75, 216; P. 71], it was concluded that small pores predominate in the porous structure of the carriers presented in Table 2.1. The regulation of porous characteristics was carried out using nitric acid peptization, as well as a number of technological techniques and additives that specifically affect the texture of alumina-oxide systems (Table 2.2). In Table 2.2 and hereinafter the following designations are adopted: GOA-K-coarse-crystalline aluminum hydroxide (bemite with crystal sizes more than 70Å with admixture of bayerite and gibbsite); GOA-M - with crystal sizes less than 60Å); AGOA-K - fine-crystalline aluminum hydroxide (mixture of pseudobemite and bemite with ammonia activated dried at 250°C GOA-K; AGOA-M - ammonia-activated calcined at 550-600°C; PVA - polyvinyl acetate; K-9 - polyelectrolyte (nitron fiber production waste); AOA-K - aluminum oxide obtained by calcination of ammonia-activated AGOA-K; AOA-M - aluminum oxide obtained by calcination of ammonia-activated AGOA-M. The carrier based on AGOA-M, containing 10% kaolin, prepared by peptization of the molding mass with nitric acid without modifiers had a specific surface area of 231 $m^2/g$. The pore volume in $cm^3/g$ (indicated in parentheses) with different radii (in nm) was as follows: <5(0.48); 40-60(0.18); 80-150(0.03); > 250(0.01) [217; P. 144, 216; P. 44]

Modification of the molding mass in obtaining alumina-kaolinborate carrier on the basis of GOA-M by various burnout additives (samples No. 1-4 in Table 2.2), did not lead to a significant change in the ratio of large and small pores, in contrast to the work [219; P. 48-52], where aluminum hydroxide was used, close in properties to sample No. 2 (Table 1.1).

**Table 2.2.**

**Физико-химические характеристики образцов носителей с фиксированным содержанием каолина AKF-78 (10 %) и бора (2,5 %)**

| № | Компоненты носителя | Модифи-катор | Пептизация; моль $HNO_3$ на моль $Al_2O_3$ | Механическая прочность | | Пористая структура | | | |
|---|---|---|---|---|---|---|---|---|---|
| | | | | Раздав-ливание; Кг/гран | Раскол; кГ/мм | Истирае-мость за 15 мин; % | Общий объем; $см^3/г$ | Эффектив-ный радиус пор; Å | Доля крупных пор; % |
| 1 | ГОА-М | | 0,012 | 1,9 | 2,32 | 10,82 | 0,76 | 4,2 | 3,2 |
| 2 | ГОА-М | ПВА | 0,051 | 2,9 | 3,78 | 8,23 | 0,44 | 3,8 | 7,8 |
| 3 | ГОА-М (АКНМ-4/16) | К-9 | 0,061 | 2.1 | 2,80 | 8,11 | 0,47 | 3,9 | 4,9 |
| 4 | ГОА-М | ПЭПА | 0,052 | 2,2 | 2,72 | 8,13 | 0,50 | 3,5 | 4,4 |
| 5 | ГОА-М+ГОА-К(40%) | ПВА | 0,052 | 2,1 | 2,13 | 9,36 | 0,90 | 4,8 | 8,3 |
| 6 | ГОА-М +ГОА-К (40%) | ПЭПА | 0,050 | 1,8 | 1,61 | 10,05 | 0,83 | 4,6 | 8,4 |
| 7 | ГОА-М +ГОА-К (40%) | К-9 | 0,030 | 2,2 | 2,10 | 9,52 | 0,73 | 4,9 | 8,7 |
| 8 | ГОА-М +АГОА-К (40%) | К-9 | 0,020 | 2,1 | 2,44 | 8,88 | 0,67 | 5,3 | 8,9 |
| 9 | ГОА-М + фракция АОА-К (10%) | ПЭПА | 0,008 | 3,1 | 3,14 | 9,63 | 0,70 | 6,4 | 31,7 |
| 10 | ГОА-М + АОА-М (10%) | - | 0,009 | 3,0 | 2,93 | 9,82 | 0,68 | 6,2 | 20,3 |
| 11 | АГОА-М + ГОА-К (40%) | - | 0,012 | 2,8 | 3,12 | 8,01 | 0,72 | 4,2 | 13,8 |
| 12 | ГОА-М+ АГОА-К (40%) | - | 0,014 | 3,0 | 3,15 | 9,06 | 0,68 | 6,1 | 14,6 |
| 13 | №9 с окунанием раствор ПЭПА+этанол+вода | | 0,008 | 3,2 | 4,23 | 9,02 | 0,58 | 7,3 | 31,7 |
| 14 | №9 с окунанием раствор МДЭА + вода | | 0,008 | 3,1 | 4,15 | 9,41 | 0,63 | 7,1 | 35,8 |

When modifying a mixture of GOA-M and kaolin AKF-78 with polyelectrolyte K-9 (a waste product of nitron fiber production), a strong carrier No. 3 was obtained (Table 2.2), with an increased, but still insufficient proportion of large pores. On its basis a sample of trimetallic catalyst AKNM-4/16 was obtained for further tests [220; P. 66]. Similar results were obtained by modifying the molding mass with polyvinyl acetate and polyethylene polyamine waste. The structure, size and properties of catalytically active centers formed by transition metal compounds on carriers are largely determined by the texture of the initial hydroxide, which in turn depends on its prehistory. Our attempt to design a carrier with the desired texture by mixing roughly dispersed hydrolytically stable GOA-K, and fine crystalline GOA-M led only to a slight increase in the proportion of large pores, with satisfactory mechanical strength (samples 5-7 Table 3.2) [211; P.123]. It was found that for industrial fine crystalline aluminum hydroxide (GOA-M) the optimum concentration of nitric acid 6% provides a mass pH of about 4.0. Under these conditions, a sufficient amount of aluminum hydroxo compounds are formed, causing strong contacts between particles during calcination. Decreasing the pH $\leq 3$ of GOA-K molding mass did not give the desired results even in terms of mechanical strength, due to the low solubility of large crystals of bemite, bayerite and gibsite and, therefore, the lack of peptization effect. When treating with nitric acid not the whole molding mixture, but separately GOA-M, followed by the introduction of kaolin and GOA-K at pH=4.0-3.5, the effect of adhesion between aluminum hydroxide particles of different sizes is fully realized. The share of desirable pores of the samples of the series, estimated by adsorption of resinous-asphaltene substances increases up to 8.7 % (sample No. 7, Table 2.2). On the basis of carrier No.7 the sample catalyst AKNM-4/11 was synthesized and tested [221; P.195]. Carrier No.8, very promising for hydrotreating catalysts, was obtained by introduction of 10 % of coarse-dispersed fraction (0.1-0.06 mm) of aluminum oxide into fine crystalline GOA-M, pre-peptized with nitric acid, similarly to the introduction of coarse-dispersed spent alumina-oxide carrier [218; P.44, 222; P.23]. The coarse-dispersed AOA-M fraction was prepared by grinding and sieving, extrudates obtained from the mass of GOA-M mixed with water and calcined at 600°C. Texture and reactivity of available powders of industrial aluminum hydroxides were also changed by thermochemical action of aqueous ammonia solutions (AGOA) according to the technology close to activation of pseudobemite powder (sample No.1 in Table 1.1) [223; C.10]. At a concentration of 0.1-1.0%, ammonia only acts as a peptizer that improves molding conditions and increases the mechanical strength of AGOA-M-based granules. In this case, there is no obvious change in the phase composition, porous structure and surface properties of the resulting aluminum oxide.

Concentrated solutions of ammonia in water, due to the expressed basic properties, actively affect the structure of surface layers of aluminum hydroxide particles possessing amphotericity. This is especially pronounced in the interaction of 15-25% $NH_4OH$ with pseudobemite of sample No. 1 (Table 1.1), when an increase in the volume of macropores with a radius of several thousand angstroms is observed, while the specific surface of the carrier monotonically drops from 300 (calcined sample No. 1) to 200 (calcined ammonia-activated sample No. 1) $m^2/g$

The GOA-M sample is closer to bemite than to pseudobemite by X-ray parameters (Fig. 2.1). The presence in the diffractograms of clear diffraction maxima from large-crystalline modifications unambiguously indicates a large proportion of aluminum trihydroxides - bayerite (d=4.71; 4.37; 3.21; 2.224 Å) and gibbsite (with d=4.83; 4.35; 4.28; 3.31; 3.16; 2.43; 2.37 Å). The large-crystalline trihydroxides are particularly pronounced in the GOA-K sample. Apparently, this brand of hydroxide is more difficult to be exposed to concentrated ammonia solution (Fig. 2.2), since all diffraction maxima of large-crystalline modifications are preserved in the X-ray diffraction pattern, while the intensity of lines from bayerite and gibbsite impurities even slightly increases. A moderately intense halo in the diffractogram caused by the formation of a new amorphous phase indicates that the interaction of aluminum hydroxide GOA-M with concentrated ammonia solution is not limited to the particle surface. Amorphization extends, at least, to the near-surface layers of the crystal lattice of weakly oxidized bemite. A slight decrease in the volume of mesopores from 0.40 to 0.32 $g/cm^3$ also testifies to the change in the dispersibility of the system. Depending on the pH of the medium, and the degree of oxstallization of the amphoteric bemite, the reactions $AlO(OH) + NH_4^+ + OH^- + H_2O \leftrightarrow NH_4^+ + Al(OH)_4^{(-)}$ and $AlO(OH) + 2NH_4^+ + 2OH^- + H_2O \leftrightarrow 2NH_4^+ + Al(OH)_5^{2-}$ occur. The pH of the GOA-M mass changes from 12.6 to 12.0 during treatment with ammonia solution. That is, due to not too strong basic properties of ammonia and increased crystallinity of bemite, compared to pseudobemite

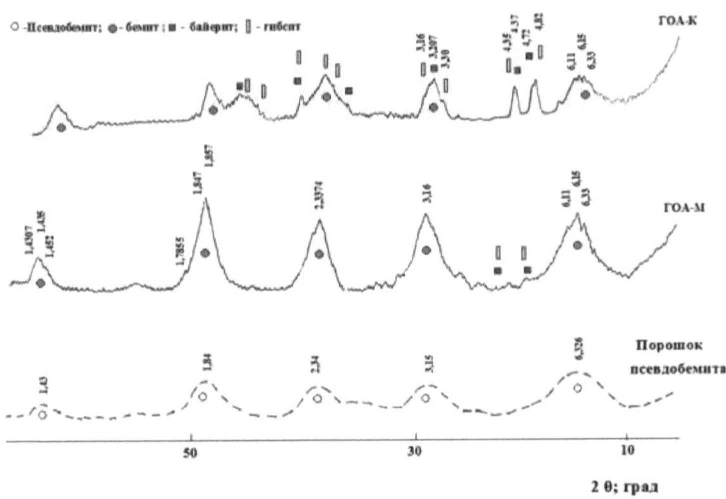


**Fig. 2.1. Diffractograms of different grades of GOA.**

the formation of aluminate anions is limited to the surface layers of particles. At the same time, intermediate products are formed in them, loosening the crystal lattice. The loosening of surface layers, even with dilute ammonia solution, reduces abrasiveness of bemite particles, increases their plastic properties and reactivity. This is what makes it possible to obtain strong extrudates on the basis of ammonia-activated GOA-M and dried at 250ºC (hereinafter AGOA-M) - carrier No. 11 (Table 2.2). Low solubility of surface layers of bemite particles in GOA-K and the presence of aluminum trihydrates, practically insoluble under the conditions of this experiment, further complicates the process of activation by ammonia. The pH value of GOA-K mass mixed with aqueous ammonia solution decreases insignificantly from 12.6 to 12.4 in the course of exposure during a day.

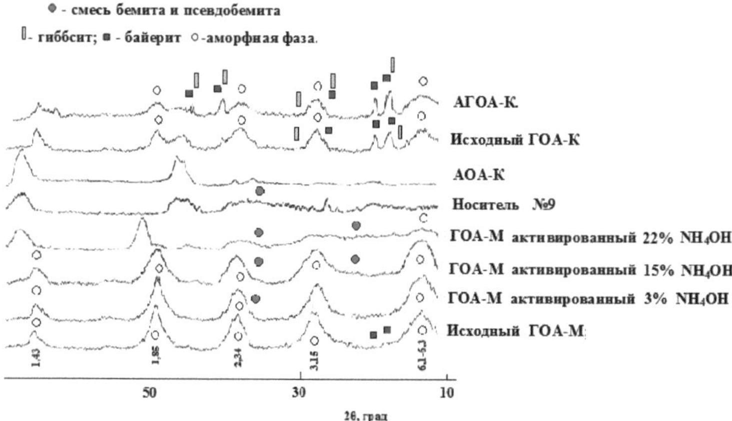


**Fig. 2.2. Effect of activation of GOA with ammonia solution on the phase composition of aluminum hydroxides.**

The sharper reflections from bayerite and gibbsite in X-ray radiographs, as compared to the unactivated samples, are explained by the ordering of their crystal structure and enlargement of crystals due to the layering of aluminate ions from the liquid phase. To some extent it is akin to aging of the precipitate of aluminum hydroxide in the mother liquor in the process of industrial synthesis of redeposited pseudobemite. The GOA-K obtained by ammonia treatment, like the activated product dried at 250°C (AGOA-K), is not molded into pellets by extrusion. The reactivity of products of destruction of near-surface layers of the crystal lattice of bemite is incomparably higher than that of already formed crystals. The combination of aluminum hydroxides of different crystallinity degree - GOA-M and activated dried AGOA-K, due to the increase of solubility in nitric acid with the formation of a mixed gel of aluminum hydroxonitrates and aluminoborates, allowed more than doubling the number of desirable large pores with good mechanical strength (sample No. 12, Table 2.2).

In the process of development of carrier No. 9, we sought to create a biporous structure (10-20 nm and 100-150 nm) by another scheme, namely, by bonding together hydrolytically stable aggregates of aluminum oxide AOA-K (fraction 0.1-0.06 mm) and particles of peptized GOA-M. At calcination, pre-treated with ammonia GOA-K, particles of activated aluminum oxide (AOA-K) with increased reactivity and larger pores than in the case of GOA-M are formed. It should be noted that the pore size and specific surface area of the obtained AGOA-K and AOA-K can be regulated by the time of exposure in water-ammonia medium and by changing the pH of the mass, as well as by the temperature of calcination. This method of preparation of alumina-kaolinborate carrier allowed to increase the share of pores available for large molecules of resinous asphaltene substances up to 31%. Samples of catalysts AKNM-3/5 and ANM-2/3 were prepared and tested on this carrier

According to the data of mercury porometry (Fig. 2.3) in aluminum oxide obtained by calcination of unactivated GOA-K, small pores predominate with a maximum on the curves of pore size distribution by radii of about 4-5 nm. In the case of GOA-M, the pore maximum is located between 4.5-5.5 nm. AOA-K, an aluminum oxide obtained by calcination of ammonia-activated AGOA-K, has a biporous structure, which is characterized by the presence of two peaks on the pore radius distribution graph: at 7-10 nm and 50-170 nm, as well as a rather weak peak at 250-500 nm. Similar results are obtained during heat treatment of activated AGOA-M, where groups of pores less than 4 nm are clearly distinguished.

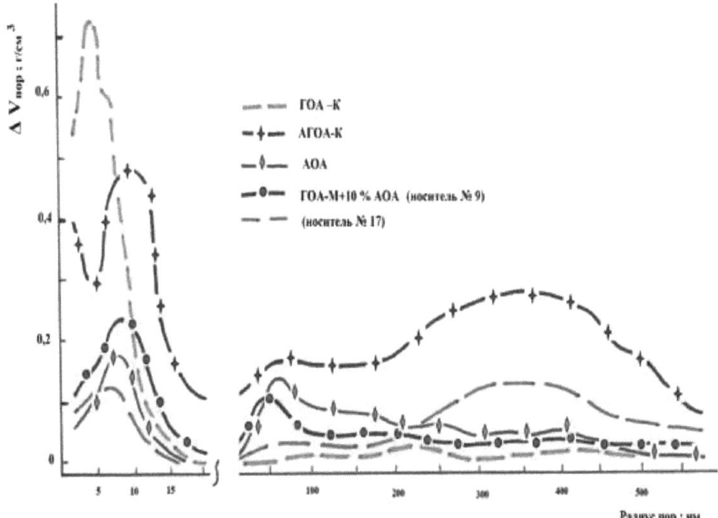


**Fig. 2.3. Effect of ammonia activation on porous characteristics of aluminum oxides and carrier based on them.**

In the process of preparation of carrier No. 9, the gel formed from aluminum hydroxonitrates and hydrated aluminoborates on the surface of GOA-M particles firmly glues large porous aluminum oxide particles and creates a secondary system of large pores partially filled with finely dispersed phase. The maximum on the pore radius distribution curves is for pores with radius 30-50 nm. The greatest influence on the growth of the volume of large pores (radius more than 10 nm) is the increase in the content of the coarse-disperse fraction of activated aluminum oxide in the composition of the carrier molding mass

From the comparison of specific surface area, total pore volume and data on the porous structure of carriers it follows that partial replacement of intermicellar liquid by dipping molded granules in the solution of texture modifier before wilting, also leads to an increase in the size of pores on the granule surface. The effect is maximized by dipping the wet pellets in a mixture of polyethylene polyamine, ethanol, water, and (1:1:1). In particular, the pore radius on the outer surface of the carrier pellets #13 increases to 180-230 Å, with an average effective pore radius of 70 Å. The presence of open large pores in the surface layer, combined with high mechanical strength and sorption activity with respect to resinous asphaltene substances, allowed us to recommend this carrier for testing as a demetallization catalyst (AKNM-5/16) [224; P. 72].

In order to implement the proposed technology and optimize the technological parameters in industrial conditions, additional equipment necessary for thermochemical activation of aluminum hydroxide and replacement of the intermicellar liquid of the formed granules was installed at the OEP of UzKFITI. In Figure 3.4 it is highlighted in red color. Using a mini-mixer, optimization of the ratio of fine crystalline kaolin AKF-78 and aluminum oxide obtained from ammonia-activated coarse-crystalline aluminum hydroxide - GOA-M: K: AOA-K was carried out (Table 2.3). Technological parameters: optimal concentration and amount of activating agent, time and temperature of ripening of activated aluminum hydroxide, and time of granules maturation in the solution of texture modifier are reflected in the regulation (Appendix). Synthesis of experimental batches of samples was carried out according to the technology of obtaining carrier No. 9 (Table 2.2) at a fixed content of modifier - boric acid (2.5 wt.%). $B_2O_3$ ). Texture and strength characteristics of carriers obtained by the proposed scheme on industrial equipment are presented in Table 2.

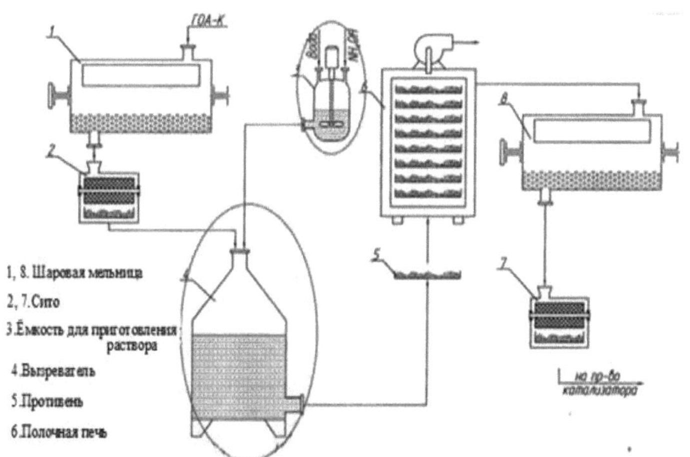


**Fig. 2.4. Technological scheme of GOA-K activation unit.**

**Table 2.3.**

**Effect of solid component ratio on media texture.**

| media no. | Weighted ratio GOA-M:K:AOA-K | $K_{strength}$; kg/mm | $D_{cp}$ of pores; nm | Share of pores with $D_{cp}$=7-10 nm; % | $\sum Vp$or; $cm^3/g$ | $S_{ood.}$; $g/m^2$ |
|---|---|---|---|---|---|---|
| №3 | 1.0:0.1:0 | 2.32 | 4.2 | 3.2 | 0.64 | 230 |
| №9 | 1,0:0,1:0,1 | 3.2 | 7.3 | 34.7 | 0.52 | 240 |
| № 15 | 1:0.2:0.3 | 2.37 | 7.0 | 31.2 | 0.52 | 250 |
| №16 | 1,0:0,5:0,1 | 3.58 | 5.4 | 18.9 | 0.32 | 110 |

| №17 | 1,0:0,1:0,5 | 1.07 | 6.6 | 30.7 | 0.67 | 280 |
|---|---|---|---|---|---|---|
| №18 | 1,0:0,2:0,5 | 1.38 | 6.2 | 27.4 | 0.58 | 220 |
| №19 | 1,0:0,5:0,5 | 1.98 | 5.1 | 21.0 | 0.50 | 140 |
| №20 | 9:0:1 | 1.2 | 4.8 | 22.6 | 0.45 | 290 |

Analysis of the influence of the ratio GOA-M: K:AOA-K on the characteristics of catalysts (Table 2.3) shows that the introduction of coarse-dispersed fraction of aluminum oxide (from 0.04 to 0.1 mm), simultaneously with kaolin improves the strength of the carrier granules. In the absence of kaolin, the necessary strength of aluminum hydroxide granules by the proposed method is not provided. The strength coefficient of carrier No. 20 was only 1.2 kg/mm. At transition to samples with high content of AOA-K with the ratio of GOA-M: K: AOA-K = 1.0:0.1:0.5 (carrier № 17) mechanical strength did not decrease as well as at the ratio of 1.0:0.2:0.5 (carrier № 18). Some increase in strength was observed at an equal ratio of kaolin to AOA-K (carrier #19), but acceptable strength was achieved either in the absence of AOA-K (carrier #1) or when the AOA-K content was reduced to a ratio of GOA-M: K: AOA-K = 1.0:0.2:0.3 (carrier #15). Increasing kaolin content at minimum AOA-K content dramatically increased strength, but equally dramatically worsened texture (carrier #16). The optimum ratio of strength and textural characteristics was achieved in the case of carriers No. 9 and No. 15.

Thus, the technological method of obtaining a bidisperse porous structure of the carrier, is to select the optimal combination of fine and coarse agglomerates (particles) in the molding mixture consisting of aluminum hydroxide, kaolin (particle size from 0.004 to 0.04 mm) with aluminum oxide (particle size from 0.04 to 0.1 mm), as well as the use of modifiers.

## §2.2 Surface properties of carriers

The surface of initial GOA-M possesses a set of moderate and very weak acid centers with $+3.8 \leq pK_a < +6.8$. Large-crystalline aluminum hydroxide (GOA-K), consisting of byerite (20÷25 nm), hydrargyllite (28÷300 nm) and bemite (20÷100 nm), has a more pronounced basic character. On its surface identified mainly very weak acid centers with $pK_a$ in the narrow range +6 ÷ +7 and some number of basic centers with $pK_a$= +9.3[217, P. 141]. On the surface of kaolin without preliminary heat treatment are present weakly acidic and weakly basic centers. After thermochemical activation of GOA-K, basic centers with $pK_{(a)} \geq +9$ were also found on its surface [217, P.141, 225; P.108]. On the surface of the prepared carrier, predominantly acidic centers are observed, the strength of which reaches $pK_a$= -5.6, and the maximum concentration falls on acidic centers of medium strength in the range of $pK_a$ from -3 to +1.5. When AOA- K is obtained, additional defects arise during the process of intensive ammonia removal, as evidenced by a sharp increase in the number of Lewis acid centers - off-lattice three-coordinated aluminum ions. The increase in the concentration of aproton centers during calcination of the samples was recorded by two independent methods of investigation. pH metric method revealed that the acidification of water in contact with AOA-K solution, indicating the presence of electron-acceptor Lewis acid centers, increases with decreasing degree of oxstallization of the initial aluminum hydroxide sample. Thus, the number of aproton Lewis centers per unit surface, increases with increasing temperature of calcination, passes through a maximum at a temperature exceeding 600°C, and then slightly decreases (Fig. 2.5) [226, P. 137-138].

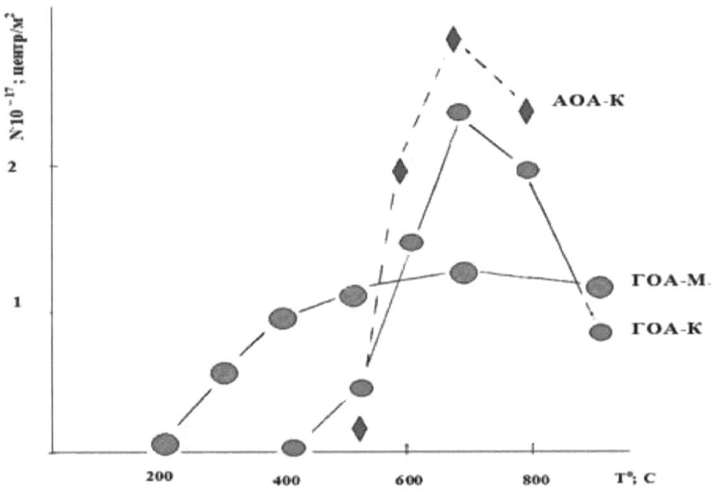


**Figure 2.5. Change of aprotonic acidity of aluminum hydroxide samples during calcination.**

Pre-activation of aluminum hydroxide with aqueous ammonia solution almost doubles the density of surface defects. The maximum concentration of Lewis acidic centers (mg-eq/m$^{2)}$, estimated from the decrease in the intensity of the dicinthalacetone absorption band at 475 nm, specific for adsorption on aprotonic centers, after dosed poisoning of acidic centers with p-butylamine, also falls at the calcination temperature of 700°C. The redistribution of the strength and concentration of acid-base centers, depending on the conditions of activation and heat treatment, is clearly shown in the temperature range 450-550°C. That is, when ammonia is almost completely removed, causing the presence of basic centers, recorded by the change in the intensity of the absorption band with a maximum at 560 nm, corresponding to the main form of indicators of electronic spectra of adsorbed phenolphthalein (Fig. 2.6).

Comparison of vibrational spectra of functional groups in the overtone region on the surface of air-dry samples and calcined at the temperature of carrier synthesis 550°C, when the decomposition process has already been completed and the transition of aluminum hydroxide into oxide has not occurred completely, revealed a difference in the concentration of surface hydroxyl groups, molecular water and residual ammonia. On the surface of the initial and calcined not modified aluminum hydroxide only hydroxyl groups in the form of overtone bands at 5.18-5.27 $kcm^{-1}$ and molecularly adsorbed water at 7.12-7.23x$cm^{-1}$ are shown (Fig. 2.7). When activated with concentrated ammonia, bands at 5.02 and 6.6 $kcm^{-1}$ appear, clearly indicating the presence of groups with N - H bonds in the composition of firmly chemisorbed ammonia, and their intensity is much less on GOA-K. It is the residual chemisorbed ammonia that is responsible for the main centers observed on the surface of activated aluminum hydroxide (Figs. 2.4, 2.5, 2.6). From the analysis of differential distribution curves of acid-base centers by strength and concentration (Fig. 2.8), it follows that ammonia activation is also accompanied by poisoning of strong and moderate acid centers. On the surface of AGOA-K, predominantly weak acid centers are formed, the strength of which is in the range $pK_a$ + ÷ +4. The maximum concentration of acid centers is located in the wide interval $pK_a$0÷+7, however, rather strong acid centers with $pK_a$ about-5 are also preserved (Fig. 2.8). Processes involving hydrogen, including catalytic hydrodesulfurization, can be influenced not only by acid-base but also by redox centers [227, P.94]. Concentrations of oxidation and reduction centers of the carrier surface were determined by the molecular probe method, after preliminary calcination at 400°C in air or in hydrogen according to the method developed earlier in UzKFITI. Diphenylamine (DFA) and anthracene (AN) molecules were used as probes for oxidation centers, which have low ionization potentials and, when interacting with acceptor centers of the surface, transfer an electron to them, transforming into the corresponding cation radicals.

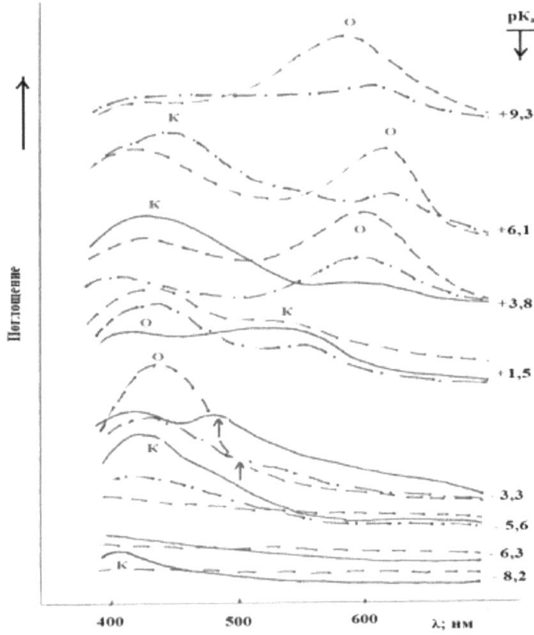


**Fig. 2.6. Diffuse reflectance spectra of indicators with different pK$_a$ adsorbed on the surface of aluminum oxides obtained by calcination at 550°C: initial ( —— ), activated by aqueous ammonia solution GOA-M ( --- ), GOA-K (–·–·– ).**

Similarly, molecules of chloranil (CA), 1,3,5-trinitrobenzene (TNB) and μ-dinitrobenzene (DNB), having high electron affinity and transforming into corresponding anion-radicals upon interaction with donor centers, were used for probing the reducing centers. The concentration of cation- and anion-radicals formed on the carrier surface during adsorption of the corresponding indicators was determined by EPR method. We found that in the process of formation of carrier No. 9, including the stage of thermochemical activation of GOA-K with ammonia, the concentration of oxidation centers decreased compared to calcined GOA-M and GOA-K, and was lower than that of carrier No. 2 obtained from GOA-M [227, P.96].

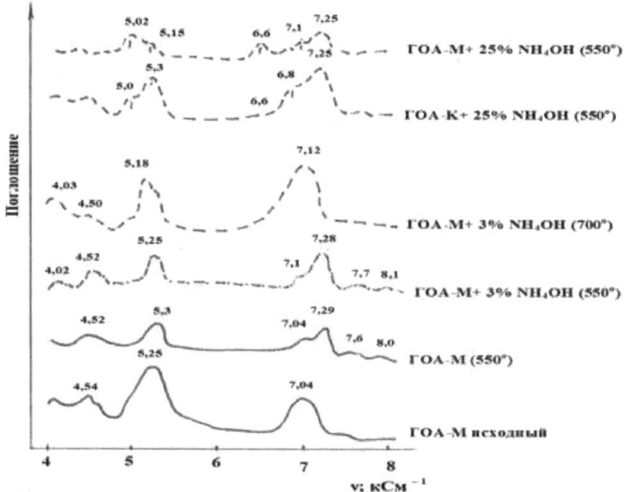


**Fig. 2.7. Change of electronic spectra of diffuse reflectance in the overtone region of the spectrum during activation of aluminum hydroxide.**

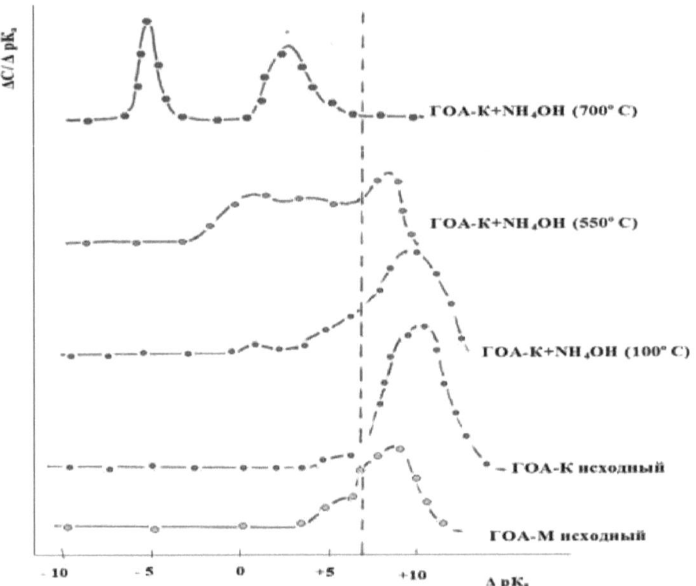


**Figure 2.8. Differential distribution curves of acid-base centers by concentration force on the surface of aluminum hydroxide, initial and after activation with ammonia solution.**

On the surface of ammonia-activated AOA-K oxidation centers were not detected, but the maximum concentration of reducing centers was observed, including strong ones, i.e. capable of forming anion radicals when interacting with μ-dinitrobenzene molecule. Heat treatment of all the studied samples in a reducing hydrogen medium was accompanied by the complete disappearance of oxidation centers and an increase in the concentration of reducing centers on the surface of both the initial components and the prepared carriers (Table 2.4).

**Table 2.4.**

**Redox properties of calcined carrier components as probed by different indicators.**

| Cipher | Center concentration; Spin/g · $10^{(-18)}$ | | | | | | |
|---|---|---|---|---|---|---|---|
| | Air-hardened | | | | | Reduced in hydrogen | |
| | Oxidative | | | Restorative | | Restorative | |
| | DF | NA | HU | TN | DNB | TN | DNB |
| GOA-M | 44, | 25, | 5,2 | 3,6 | 0,08 | 17,5 | 1,5 |
| GOA-C | 32, | 14, | 1,3 | 0,6 | * | 3,8 | 0,07 |
| AGOA | - | - | - | 4,5 | 0,19 | 11,6 | 2,3 |
| Kaolin | 1,1 | - | - | - | - | * | - |
| Carrier #2 | 3,6 | 2,2 | 0,3 | 1,3 | 0,09 | 1,8 | 0,12 |
| Carrier #9 | 2,7 | 1,4 | 0,2 | 1,5 | 0,17 | 2,3 | 0,53 |

Thus, by varying the fractional composition of long-term storage aluminum hydroxide powders, conditions of their thermochemical activation and modification, the technology of synthesis of carriers for hydrotreating catalysts based on local raw materials and non-liquid aluminum hydroxide has been developed. The carriers prepared according to the developed technology are characterized by bimodal distribution of pores by radii, a large proportion of mesopores, and a favorable spectrum of surface centers.

# CHAPTER III. CATALYTIC PROPERTIES OF TRIMETALLIC CATALYST

## §3.1 Degrees of transformation of model substances on ACNM catalysts under low hydrogen pressure conditions

The primary evaluation of catalytic activity and selectivity of the developed catalysts was studied on the example of model reactions of hydrogenolysis of dioctyldisulfide (light sulfur) and thiophene (more difficult sulfur). Conversion of sulfur compounds was carried out on a flowing microcatalytic unit at a small excess pressure of hydrogen. The activity and selectivity of the studied catalysts, calculated from the results of chromatographic analysis, are presented in Table 3.1. The aggregate of the obtained data on the activity of a series of catalysts and the content of structures differing in solubility in water and aqueous ammonia allows us to come to the conclusion about the existence of different types of active centers on the surface of catalysts. As it was mentioned above, the direct hydrodesulfination reaction with C-S bond breaking is catalyzed by structures soluble in aqueous ammonia solution, including those extracted from the catalyst during water treatment. The correlation between the amount of selectively extracted molybdenum ions, as well as elements of group VIII of the Periodic System and catalytic properties is clearly seen in Table 3.1. Conversion of thiophene and dioctyldisulfide proceeds to some extent even on the catalyst AKNM-5/16-P on a wide-porous carrier #13, recommended for use as a protective layer catalyst. AKNM-5/16-P catalyst contains only 5 wt.% of MoO(3), water extracted with water.

$MoO_3$, water extracted only 0.8 wt. % of the initial molybdenum content in the catalyst, and water-ammonia solution - 1.8 wt. %. In the works of Spozhakina and other authors it was noted that molybdenum compounds are not extracted at all when treated with water and/or ammonia-water solution at their content below 6-7 wt. % of MoO(3). $MoO_3$. This difference, in our opinion, is due to the presence of silicon compounds in the composition of carriers, which interact with ammonium paramolybdate during the synthesis of catalysts with the formation of some amount of water-soluble silicon-molybdenum heteropolyacid. This conclusion was confirmed by qualitative reaction during the analysis of extracts from AM-P and #1-AM model catalysts. Extracts from the catalysts after addition of ascorbic acid acquired blue coloration characteristic of $Mo^{5+}$ ions in silicon heteropoly compounds. The color intensity of the aqueous extract from the AM of the catalyst containing 9 wt.% MoO(3) on the carrier. $MoO_3$ on the carrier No. 16 with high kaolin content (ratio of GOA-M: K: AOA-K=1.0:0.5:0.1), was much brighter. Consequently, water-soluble silicon molybdenum compounds were formed much more during the synthesis of the model sample - about 10 % of the initial molybdenum content. From the low-percentage trimetallic catalyst AKNM - 5/16 a very small amount of silicon-molybdenum compounds was extracted, along with promoting transition metals of group VIII. The amount of transition metals extracted by aqueous ammonia solution, i.e. potentially active in hydrodesulfination, is not more than 0.13 % of the weight of the catalyst. This value increases with increasing transition metal content.

**Table 3.1.**

**Результаты испытания образцов катализаторов в конверсии тиофена, диоктилдисульфида на микрокаталитической установке.**

**(давление водорода - 0,001 МПа, водород:сырье = 20:1, объемная скорость 0,5 час$^{-1}$, фракция 0,63-1,0 мм, навеска катализатора – 1г)**

| Катализатор | Содержание, % | | | Конверсия, % | | | Состав продуктов превращения при 300°С, % | | | | | |
|---|---|---|---|---|---|---|---|---|---|---|---|---|
| | | | | Тиофен | | Диоктилдисульфид (300°С) | Тиофен | | | Диоктилдисульфид | | |
| | $MoO_3$ | CoO | NiO | 300°С | 400°С | | $\Sigma C_2+C_3$ | н-$C_4H_{10}$ | $C_nH_{2n}$ | $\Sigma C_1 \div C_7$ | н-$C_8H_{18}$ | $\Sigma C_nH_{2n}$ |
| АКНМ-5/16 | 5,0 | 0,49 | 0,48 | 3,8 | 6,4 | 9,6 | 0,3 | 6,6 | 76,8 | 1,8 | 11,6 | 68,4 |
| АКНМ-5/16 * | 4,96 | 0,47 | 0,47 | 3,6 | 6,3 | 8,9 | 0,3 | 6,0 | 77,4 | 1,7 | 11,6 | 68,3 |
| АКНМ-5/16** | 4,91 | 0,47 | 0,46 | 2,7 | 6,3 | 7,1 | - | 5,8 | 77,6 | 0,3 | 11,2 | 68,8 |
| АКНМ-3/5 | 11,91 | 2,51 | 1,50 | 21,9 | 26,5 | 43,5 | 0,8 | 29,7 | 49,3 | 2,2 | 51,2 | 28,7 |
| АКНМ-3/5* | 10,56 | 2,05 | 1,38 | 18,1 | 23,2 | 34,2 | 0,8 | 21,8 | 57,6 | 1,9 | 38,2 | 41,9 |
| АКНМ-3/5** | 6,44 | 1,86 | 1,30 | 4,2 | 7,1 | 10,4 | 0,5 | 7,27 | 71,7 | 0,8 | 12,9 | 67,0 |
| АКНМ-2/9 | 17,5 | 4.19 | 1.70 | 65,8 | | 90,3 | 1,4 | 95.6 | 2.1 | 3,3 | 58,4 | 37,9 |
| АКНМ-2/9* | 11,9 | 1,95 | 0,96 | 37,1 | | 44,8 | 0,8 | 32,2 | 66,3 | 2,4 | 16,2 | 80,8 |
| АКНМ-2/9** | 9,83 | 1,28 | 0,42 | 23,4 | | 15,8 | 0,6 | 31,2 | 67,4 | 1,8 | 13,6 | 83,9 |
| *АНМ-2/3* | 16,2 | 0 | 5,01 | 27,8 | 39,7 | 56,6 | 1,3 | 97,8 | 2,2 | 3,4 | 65,2 | 3,8 |
| *АНМ* | 11,5 | 0 | 2,73 | 20,2 | 23,4 | 42,2 | 0,6 | 24,5 | 51,2 | 2,3 | 44,6 | 27,2 |
| *АКМ* | 11,8 | 3,97 | 0 | 22,0 | 26,5 | 43,3 | 0,4 | 29,8 | 48,5 | 2,8 | 51,6 | 28,2 |
| *АКМ (пром)* | 12,0 | 3,93 | 0 | 21,8 | 26,7 | 43,4 | 0,9 | 29,1 | 49,1 | 2,3 | 51,6 | 28,9 |

In the catalyst AKNM-3/5 after treatment with aqueous ammonia solution, the sum of molybdenum, cobalt and nickel oxides decreased from 15.91 to 9.6; wt. and amounted to about 6% of the catalyst weight. For the catalyst AKNM-2/9, oriented on hydrogenation of polyaromatic compounds, the sum of active components potentially active in C-S bond breaking reactions reached almost 14% of the catalyst weight.

The degree of conversion of thiophene and dioctyldisulfide was taken as a measure of hydrodesulfurization activity, estimated by the decrease in the area of the corresponding peaks in the chromatograms after the reaction. The flow rate of thiophene fed to the microcatalytic unit was 2.38 μmol/hour per gram of catalyst, and that of dioctyldisulfide was 94.3 μmol/hour (Table 3.1). The fraction of n-butane (n-octane) identified among the products formed was used as an estimate of hydrogenation activity. The cracking ability was judged by the sum of $C_2$-$C_3$ hydrocarbons; the presence of methane was not recorded. Under our chosen conditions of model processes on the microcatalytic unit, the degree of thiophene conversion increased accordingly, from 21.9 % on the catalyst AKNM-3/5 to 65.8 % on AKNM-2/9. Among bimetallic catalysts with close content of transition metals the maximum conversion of thiophene was achieved on AKM catalyst. And among the studied trimetallic samples, AKNM-2/9 catalyst obtained by single impregnation with a joint solution stabilized with phosphoric and citric acids prevailed. Due to this technological method, the AKNM-2/9 catalyst contained 17.5 wt.% of MoO(3) and only 3.5 wt.% of MoO(3). $MoO_3$ and only 3.8 wt. % of $P_{(2)}O_{(5)}$. $P_2O_5$. The catalyst ANM-2/3-

P, containing a close amount of active metals but obtained by a single impregnation without the addition of citric acid, contained about 7.6 wt. % of wt. $P_2O_5$. Trimetallic catalysts containing about 12 % $MoO_3$ on carrier #9, as well as other studied carriers, occupy an intermediate position. In Figure 3.1, the dependence of activity, in micromoles of sulfur compound, referred to the total transition metal content (TMC), reacted on each of the catalysts on the same carrier, a close ratio of molybdenum, cobalt and nickel, but with a different total concentration of active components, is clearly seen

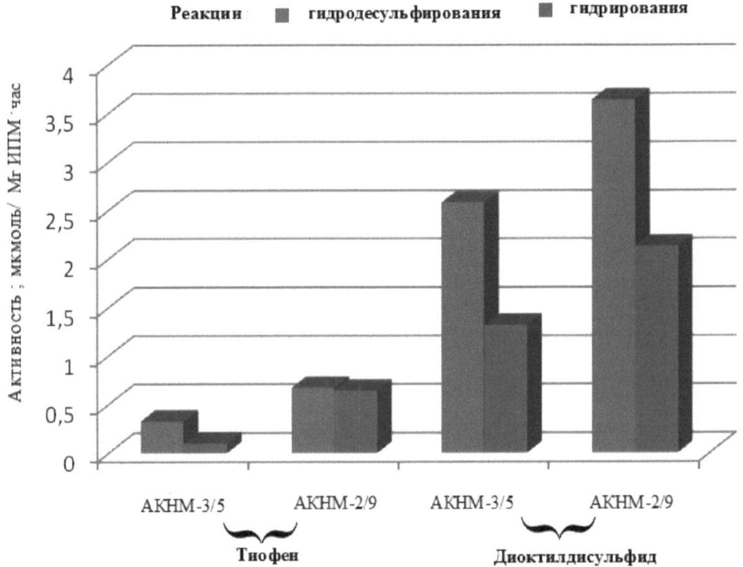


**Fig. 3.1. Dependence of catalyst activity in the reactions of conversion of sulfur compounds and subsequent hydrogenation of conversion products on the total concentration of active metals.**

Although with increasing concentration of IPM an increase in activity was observed both in the conversion of model compounds and subsequent hydrogenation of unsaturated reaction products, still the increase in hydrogenation activity clearly outpaced the increase in the amount of reacted substances.

In literature sources, this phenomenon is most often associated a progressive increase in the number of molybdenum structures very weakly bound to the carrier surface. Calculations made on the basis of the results of chromatographic analysis of initial substances and products of their hydrotreatment at the microcatalytic unit (Table 3.1) are similarly [84; P.177] presented in Tables 3.2 and 3.3. In Tables 3.2-3.3 the results of experiments with catalysts after extraction of weakly bound (physically adsorbed) transition metal compounds with water (WE) are highlighted with yellow background, followed by extraction with water-ammonia solution of IPM compounds bound to the surface by chemical bonds (AVE) - with orange background, and treatment with water-ammonia solution only - with green background. Treatment of the initial catalysts with water-ammonia solution removes both weakly bound and more strongly bound structures of IPM (VE+ABE). The amount of transition metals (in terms of the corresponding oxides) in the initial catalysts, as well as those remaining after the extraction stage (i.e., very strongly bound, embedded in the carrier volume, IPM compounds that are not removed by extraction, under the selected conditions) was determined by a combination of electron-probe and chemical analysis of representative samples after their repeated heat treatment at 400°C. The amount of transition metals removed by extraction was determined from the difference of the obtained results by calculation.

**Table 3.2.**

**Comparison of catalyst activity in model reactions of thiophene hydrodesulfination and hydrogenation of unsaturated thiophene hydrodesulfination products**

| | ∑ IPM (Mo, CoNi); mg/g of catalyst | Amount of thiophene | Activity in C-S bond breaking reactions; µmol | Amount of butane; µmol/h | Activity in hydrogenation of conversion products |
|---|---|---|---|---|---|
| AKNM-5/16 | | | | | |
| Initial | 59,7 | 9,04 | 0,15 | 0,60 | 0,01 |
| After the VE. | 59,0 | 8,57 | 0,14 | 0,51 | 0,009 |
| After VE and AVE | 50,0 | 6,43 | 0,13 | 0,37 | 0,007 |
| VE deleted. | 0,7 | 0,47 | 0,67 | 0,09 | 0,13 |
| Deleted VE+ABE | 9,7 | 2,61 | 0,27 | 0,23 | 0,02 |
| Removed by AVE | 9,0 | 2,14 | 0,24 | 0,14 | 0,015 |
| AKNM-3/5 | | | | | |
| Initial | 159,2 | 52,1 | 0,327 | 15,5 | 0,097 |
| After the VE. | 139,9 | 43,08 | 0,308 | 9,39 | 0,067 |
| After VE and AVE | 96,0 | 9,996 | 0,104 | 0,73 | 0,008 |
| VE deleted. | 19,3 | 9,02 | 0,47 | 6,11 | 0,32 |
| Deleted VE+ABE | 63,2 | 42,1 | 0,67 | 14,77 | 0,23 |
| Removed by AVE | 43,9 | 33,08 | 0,75 | 8,66 | 0,20 |
| AKNM-2/9 | | | | | |
| Initial | 233,9 | 156,6 | 0,67 | 149,7 | 0,64 |
| After the VE. | 148,1 | 88,3 | 0,60 | 28,4 | 0,19 |

| | | | | | |
|---|---|---|---|---|---|
| After VE and AVE | 115,3 | 55,7 | 0,48 | 17,4 | 0,15 |
| VE deleted. | 85,8 | 68,3 | 0,80 | 121,3 | 1,41 |
| Removed VE+ABE | 118,6 | 100,9 | 0,85 | 132,3 | 1,11 |
| Removed by AVE | 32,8 | 32,6 | 0,99 | 11,0 | 0,33 |

**Table 3.3.**

**Comparison of catalyst activity in model reactions of dioctyldisulfide hydrodesulfination (DOS) and hydrogenation of unsaturated products of dioctyldisulfide hydrodesulfination**

| | ∑ IPM (Mo, CoNi); | Quantity of DOS converted; kmol/hour | Activity in C-S bond breaking reactions; μmol DOS/mg | Amount of octane; μmol/h | Activity in hydrogenation of DOS conversion products; μmol octane/mg IPM·hour |
|---|---|---|---|---|---|
| AKNM-5/16 | | | | | |
| Initial | 59,7 | 90,5 | 1,52 | 10,5 | 0,18 |
| After the VE. | 59,0 | 83,9 | 1,42 | 9,7 | 0,16 |
| After VE and AVE | 50,0 | 67,0 | 1,34 | 7,5 | 0,15 |
| VE deleted. | 0,7 | 6,6 | 9,42 | 0,8 | 1,14 |
| Deleted VE+ABE | 9,7 | 23,5 | 2,42 | 3,0 | 0,3 |
| Removed by AVE | 9,0 | 16,9 | 1,87 | 2,2 | 0,24 |
| AKNM-3/5 | | | | | |
| Initial | 159,2 | 410,2 | 2,58 | 210,0 | 1,32 |

| | | | | | |
|---|---|---|---|---|---|
| After the VE. | 139,9 | 322,5 | 2,30 | 123,2 | 0,88 |
| After VE and AVE | 96,0 | 98,1 | 1,02 | 12,7 | 0,13 |
| VE deleted. | 19,3 | 87,7 | 4,54 | 86,8 | 4,49 |
| Removed VE+ABE | 63,2 | 312.1 | 4,94 | 197,3 | 3,12 |
| Removed by AVE | 43,9 | 224,4 | 5,11 | 36,4 | 0,83 |
| AKNM-2/9 | | | | | |
| Initial | 233,9 | 851,5 | 3,64 | 497,3 | 2,13 |
| After the VE. | 148,1 | 422,5 | 2,85 | 684,4 | 0,46 |
| After VE and AVE | 115,3 | 149,0 | 1,29 | 20,3 | 0,18 |
| VE deleted. | 85,8 | 429,0 | 5,0 | 428,9 | 5,0 |
| Deleted VE+ABE | 118,6 | 702,5 | 5,92 | 477,0 | 4,02 |
| Removed by AVE | 32,8 | 273,5 | 8,34 | 48,1 | 1,47 |

As follows from Tables 3.1-3.3, as transition metals were successively extracted from the catalysts with water and ammonia solution, the degree of transformation of all the studied model substances gradually decreased. The activity in hydrodesulfination and hydrogenation reactions on the samples containing only residual compounds of molybdenum and promoter elements of group VIII, not removed by extraction according to our methodology, was very low. It was shown that the level of hydrodesulfurization activity in the process of transformation of sulfur compounds correlates to a greater extent with the content of molybdenum and some amount of Co and/or Ni, which are not extracted during the treatment of catalysts with water, but are removed by subsequent extraction with aqueous ammonia solution

On the example of dioctyldisulfide conversion on AKNM-2/9 catalyst (Fig. 3.2, Table 3.2) it is most clearly seen that the maximum ability to hydrodesulfidation is possessed by structures (mainly molybdenum) sufficiently firmly bound to the carrier, due to chemical bonds, which are not removed when the catalyst is treated with water, but are well extracted by water-ammonia solution (extracted from catalysts by water-ammonia solution)

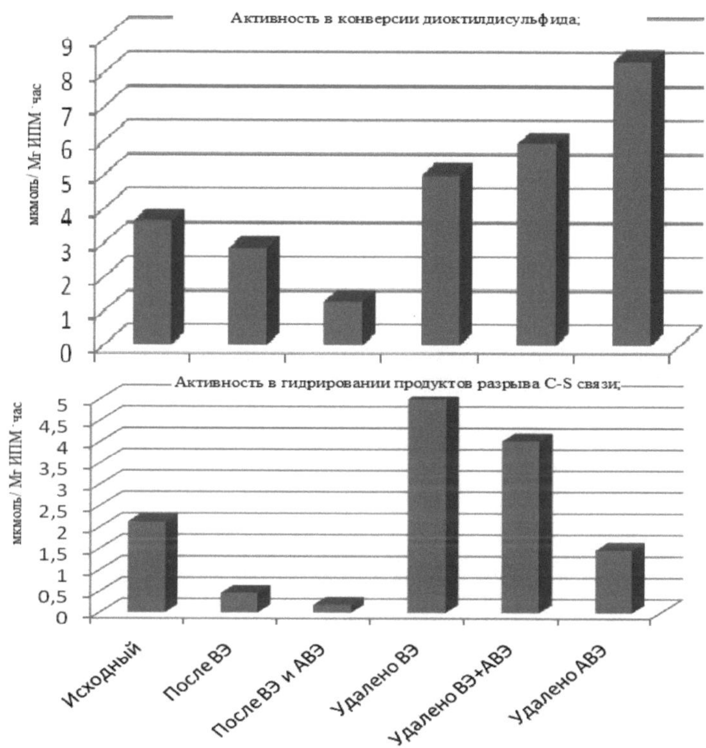


**Fig. 3.2. Comparative activity of AKNM-2/9 catalyst in hydrodesulfurization of dioctyldisulfide and**

**hydrogenation of its conversion intermediates after selective extraction of different structures with transition metal ions.**

This is well illustrated by the figures, where the activity of the catalysts is related to the total amount of transition metals contained in the samples under study. Separately presented activity in the transformation of model sulfur compounds on the initial catalyst, and preserved after the completion of sequential aqueous and water-ammonia or single extraction with aqueous ammonia solution

The minimum content of hydrocarbons with unsaturated bonds ($C_nH_{2n}$) was found in the hydrogenizate on the AKNM-2/9 catalyst, with the maximum amount of water-extractable molybdenum, cobalt, and nickel. The proportional dependence of activity on the amount of Mo, Co and Ni (water-extractable) structures weakly bound to the carrier was observed only if the concentration of $MoO_3$ in the catalyst composition was higher than 9 wt%. In particular, the maximum contribution to the hydrogenation of butenes (1.41μmol/mg IPM· hour), unsaturated intermediates thiophene conversion, was provided by 85.8 mg of water-soluble IPM in 1 gram of AKNM-2/9 catalyst. And on AKNM-3/5 and AKNM-5/16 catalysts: activities of 0.32 and 0.13 μmol/mg IPM· hour were achieved at the expense of 19.3 and 0.7 mg of water-soluble IPM, respectively. The degree of hydrogenation of cyclohexene, modeling olefins - intermediates of hydrogenolysis, increased in proportion to the total concentration of molybdenum, nickel and cobalt structures soluble in water and water-ammonia solution [245; P.46].

## §3.2 Relationships of hydrodesulfurizing and hydrogenating activity of catalysts with the number of IPM structures

Further, hydrogenating and hydrodesulfurizing activity of initial catalysts with different content of transition metals in the process of transformation of dioctyldisulfide (a compound modeling "light sulfur") and thiophene (modeling compounds of "difficult sulfur" in oil fractions, but not complicated by steric factor) was compared. Measurements of activity of initial catalysts on the pilot plant were repeated under identical conditions, but after removal of the most weakly water-bound transition metal compounds, as well as comparatively strongly bound structures The regularities revealed on the microcatalytic unit in the process of transformation of model substances were confirmed when studying the activity of the developed catalysts on the high-pressure pilot plant in the process of hydrotreating of real oil fractions (Fig.3.3)..3, Table 3.4) [244; C.40, 246;C.228,247;C.46, 248;C.75].

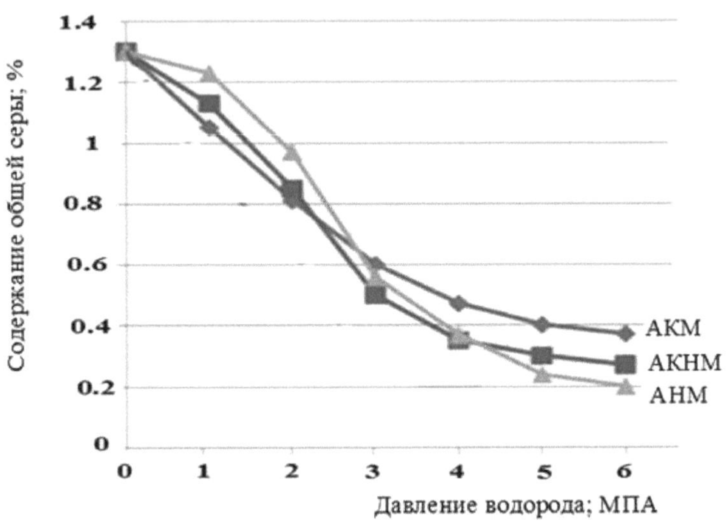


**Fig. 3.3. Dependence of total sulfur concentration reduction on hydrogen pressure during hydrotreating of deasphaltized residue on different catalysts**

Under the condition of relatively low hydrogen pressures (2MPa) and temperatures, the hydrodesulfurization activity was higher on aluminocobaltmolybdenum catalysts, both industrial and synthesized. At pressure increase up to 2.5MPa this index is equalized for AKNM-R and AKM-R samples, and at pressure 3.0-4.0 MPa in the temperature range of 300-340°C the activity of AKNM-3/5-R sample exceeds the hydrotreating level quite significantly by the main parameters: viscosity index, chromaticity, sulfur and nitrogen content. Color at purification of medium viscous oils was in the range of 2.0-2.6 units. CNT, and sulfur content decreased from 0.91 to 0.20-0.32 mass %.

The analysis of the gas phase revealed that in the reaction products, in addition to hydrogen sulfide, in all the samples studied no more than 0.2 vol. % of light hydrocarbons, mainly pentanes, and lighter hydrocarbons are absent. An indirect characteristic of the depth of transformation of heterorganic compounds and hydrocarbons of feedstock in the hydrotreating process is the change in the composition of hydrogen-containing gas. It is proved that on the considered catalysts, given in Table 3.4, there was no appreciable increase in the content of light hydrocarbons in the hydrogen-containing gas at the reactor outlet, as compared to the industrial catalyst AKM. The recorded appearance of ammonia in the reaction products indicated hydrogenolysis of nitrogenous compounds.

It is known that the reactivity of different classes of sulfur compounds is different. Therefore, for modeling the processes of removal of "easy", "moderately difficult" and "difficult" sulfur in oil fractions we obtained the following model mixtures. Conditionally we can assume that after the hydrotreating cycle of the feedstock portion, at hydrogen pressure of 3 MPa, the content of easily desulfurizable compounds in the obtained hydrogenizate (No.1) decreases almost to zero. Hydrogenizate №1 modeled oil feedstock with total sulfur content of 0.57 %, but high ratio of "difficult", spatially hindered sulfur and "moderately difficult" thiophene sulfur, i.e. natural compounds specific for feedstock of Fergana oil refinery. Hydrogenizate (№2) - deeply hydrotreated deasphaltizate (up to sulfur content 0,1 % wt.) on industrial aluminonickel-molybdenum catalyst at pressure 6MPa and temperature 340°C, served as a basis for artificial model mixtures. The content of residual "difficult sulfur" at repeated hydrotreating under the same conditions decreased by less than 0.01 % wt%, i.e. was within the accuracy of the experiment.

When dioctyldisulfide (1.3 wt.% sulfur) was introduced into hydrogenizate No. 2, an artificial model mixture with a high content of exclusively "light" disulfide sulfur was obtained. On this model mixture, the catalysts AKM-R, AKNM-R and ANM-R were tested on carrier No. 9 containing the same amount of transition metals (12.3-12.5 wt. % MoO(3)). $MoO_3$, combined with 3.9-4.1 wt% of CoO and/or NiO), in the process of dioctyldisulfide hydrodesulfination. A clear advantage of the cobalt-molybdenum catalyst in the hydrogen pressure range allowed at the G-24 unit of the Fergana Refinery was revealed. At pressure of 3 MPa the content of disulfide sulfur on AKM-R/#9 decreased to 0.052 wt. %, on AKNM-R/#9 to 0.056 wt. %, and on ANM-R/#9 only to 0.093 wt. % of conditionally "light sulfur". The above pattern was maintained at increasing hydrogen pressure up to 4 MPa. In the region of higher hydrogen pressures the best results on hydrodesulfination of dioctyldisulfide in the artificial model mixture of "light sulfur" were obtained on the catalyst ANM/#9. The decreased activity of cobalt-containing catalysts AKNM/#9 and, moreover, AKM/#9, in comparison with nickel-containing analogs does not contradict numerous experiments described in literature sources.

The detected regularity coincided quite well with the results of tests on natural raw material - deasphaltizate with high content of total sulfur - 1.3 %, presented in Figure 3.3 [249; C.11-12].

**Table 3.4.**

**Test results of catalysts in the process of hydrotreating of feedstock for base oils production.** $T = 300^oC$, $G = 1.0\ h^{-1}$, $P = 3.0\ MPa$

| Indicators | Exodus. oil | AKM GO-70 industrial. | ANM (nos.-9) | AKM (nos.-9) | AKNM-4/1( (nos.-3) | AKNM-3/5 (nos.-9) $MoO_3$-11,91 |
|---|---|---|---|---|---|---|
| Deasphaltized residue (after 1000 h under severe conditions) | | | | | | |
| Color,; units. CNT | >8,0 | 5,2/5,2 | 5,3/5,3 | 5,3/5,3 | 5,2/5,2 | 5,2/5,2 |
| Total sulfur content; % | 1,30 | 0,56/ 0,57 | 0,58/ 0,59 | 0,54/ 0,55 | 0,52/ 0,52 | 0,50/ 0,50 |
| Viscosity at 40° C; cSt | 305,6 | 303,5/ 303,6 | 315,0 / 315,0 | 280,0 / 280,0 | 280,0 / 280,0 | 312,2/ 312,2 |
| Viscosity at 100° C; cSt | 22,5 | 21,64/ 21,65 | 22,9/ 22,8 | 21,0/ 21,0 | 21,0/ 21,0 | 22,3/ 22,3 |
| Viscosity index | 91,0 | 85,0/85,0 | 90,0/ 90,2 | 87,2/ 87,2 | 88,7/ 88,7 | 90,0/90,0 |
| Density at 20° C, g/cm3 | 0,9003 | 0,9227/ 0,9229 | 0,9109/ 0,9108 | 0,9207/ 0,9207 | 0,9209/ 0,9209 | 0,9217 / 0,9217 |
| Solidification temperature,° C | -14 | -15/-15 | -14/-14 | -14/-14 | -15/-15 | -16/-16 |
| III fraction of vacuum distillate | | | | | | |
| Color,; units. CNT | 5,5 | 2,1 | 2,5 | 2,5 | 2,5 | 2,0 |
| Total sulfur content; % | 0,91 | 0,22 | 0,32 | 0,26 | 0,24 | 0,20 |
| Viscosity at 100°C; cSt | 6,97 | 8,0 | 6,8 | 6,8 | 7,0 | 6,8 |
| Viscosity index | 94,5 | 97,6 | 96,5 | 96,5 | 98,5 | 98,2 |
| Solidification temperature,° C | -17 | -18 | -16 | -16 | -18 | -19 |

Figure 3.3 also showed the improvement of process parameters when switching to the high-percentage catalyst AKNM-2/9, obtained by single impregnation of the carrier

No. 9 with a combined solution of appropriate transition metal salts stabilized with phosphoric and citric acid.

Further, by addition of thiophene (1.3 wt.% sulfur) to hydrogenisate #2 an artificial model mixture with high content of exclusively "moderately difficult" thiophene sulfur was obtained. The prepared model mixture was more fully subjected to hydrodesulfurization at relatively low pressures over AKNM/#9 catalyst. Where at hydrogen pressure of 3.0MPa the content of residual thiophene sulfur amounted to 0.12 %, and at 4 MPa it decreased to 0.08 %. On catalysts ANM/#9 and AKM/#9 the degree of thiophene sulfur conversion at hydrogen pressure of 3.0MPa was 90.1 and 89.8 % of the initial one, respectively. With pressure increase up to 4.0MPa the degree of transformation on the investigated catalysts was: AKNM/#9 - 93.8%, ANM/#9 - 92.7%, AKM/#9 - 92.0%. When the catalysts were tested under similar conditions on hydrogenisate #1-natural model mixture with a high ratio of "difficult", spatially hindered sulfur and partially removed thiophene sulfur, most likely, "moderately difficult" thiophene sulfur was more likely to undergo transformation (Fig. 3.4). Due to a sufficient amount of weakly bound structures (water-extractable nickel molybdate), hydrogenation of double bonds in thiophene rings occurred, but much more efficiently than when model thiophene was tested at the microcatalytic unit (Table 3.1)

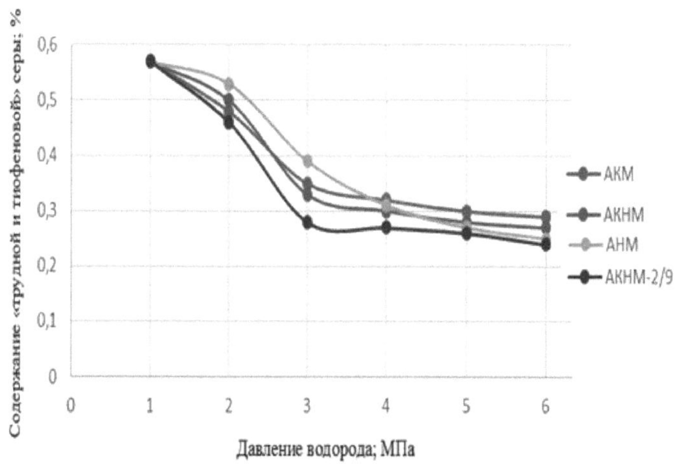


**Fig. 3.4. Dependence of "difficult and thiophene" sulfur concentration reduction on pressure in the process of hydrotreating of hydrogenizate containing 0.57 % of difficult to remove sulfur compounds.**

The simultaneous presence of a high concentration of Co-Mo structures extracted by aqueous ammonia solution provided a sufficiently effective desulfurization of hydrogenation products. At a pressure of 3.0 MPa the degree of transformation on the investigated catalysts increased in the series: AKM/$^1$9-43.8 %< ANM/$^1$9 - 61.4 % < AKNM/$^1$9-64.9 %, and when the process was carried out at 4.0 MPa the order changed: AKM/$^1$9-61.4 % < AKNM/$^1$9-70.18 %,< ANM/$^1$9-73.9 %.

This fact, in combination with the data on the conversion of thiophene sulfur in an artificial model mixture, proved that the hydrogenation of double bonds in sulfur compounds of aromatic character occurs at pressures above 4.0 MPa and requires a catalyst with a high content of weakly bound to the surface nickel-molybdate structures with maximum hydrogenation activity

Low efficiency of catalyst AKM-R/#9 was confirmed by comparative tests of catalysts on natural model mixture ( obtained by hydrotreating of high-sulfur deaphthaltic residue at hydrogen pressure of 2.5-3.0 MPa on catalyst AKNM-2/9) containing mainly "difficult sulfur" (Fig. 3.5). In this case, even at very high hydrogen pressure, the degree of conversion of "difficult sulfur" molecules at not high initial concentration (0.27% wt.) did not exceed 15%.

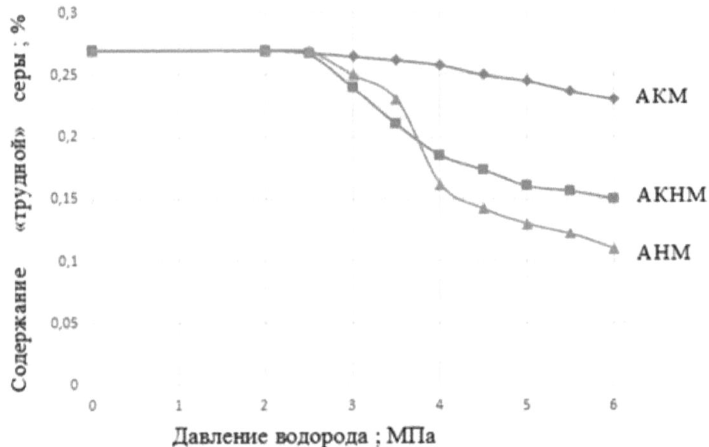


**Fig. 3.5. Dependence of "difficult" sulfur concentration reduction on pressure in the process of hydrotreating of hydrogenizate containing 0.27 % of difficult to remove sulfur compounds.**

The advantage of trimetallic AKNM-P/#9 catalyst over ANM-P/#9, as well as on real hydrocarbon feedstock (Table 3.5), was manifested only in a very narrow pressure range of 2.5-3.5 MPa. However, changing the synthesis method, simultaneously with increasing the concentration of active components (catalyst AKNM-2/9 in Table 4.2 and 4.4), allowed to reduce the sulfur content in the model mixture of "difficult sulfur" 0.21 % wt. S, and total sulfur in a typical high-sulfur feedstock to 0.23 % wt. S, even at relatively low hydrogen pressure of about 3MPa for hydrotreating processes (Table 3.5).

The data of Table 3.5 also show that in case of realization of the method of sulfidation of catalyst AKNM-2/9 by gaseous hydrogen sulfide, widely used at a number of foreign refineries, the sulfur content in base oils can be reduced to 0.5% (50ppm). Thus, to obtain highly active AKNM series catalysts, the ratio of GOA-M: K: AOA-K carrier components should be in the range from 1.0:0.1:0.1 to1.0:0.2:0.3. Although at the ratio of 1.0:0.2:0.5 with a high proportion of activated aluminum oxide in the carrier the activity of the catalysts is high, but the granules do not meet the strength criterion. An extremely important performance characteristic of catalysts, in addition to activity, is stability in the hydrotreating process. Express testing of stability under harsh conditions, often used in research laboratories, by a short-term (several hours) increase in temperature and pressure in the reactor, showed the following.

After the negative effect of high temperature during hydrotreating of unfavorable feedstock (with high content of sulfur, nitrogen and unsaturated compounds), catalysts on carrier No. 9 (with the ratio of GOA-M: K: AOA-K = 1.0:0.1:0.1) AKNM-3/5 decreased the initial activity by 0.13 %, and AKNM-2/9 by 0.33 %

The catalyst AKNM/#17 on the carrier with the ratio of 1.0:0.5:0.5 reduced the activity by 0.49%, and the catalyst AKNM/#20, not containing kaolin by 2.6%. The obtained data show that the introduction of kaolin and activated aluminum oxide by the proposed technology at the same

**Table 3.5.**

**Сопоставительные данные по каталитическим свойствам катализаторов в процессе гидрообессеривания масляных фракций, полученные на пилотной установке при объемной скорости – 1 ч$^{-1}$.**

| Шифр катализатора | Тем-пера-тура про-калки; °С | Соотно-шение ГОА-М: К:АОА-К | Сульфидирование | NiO | CoO | $MoO_3$ | | Параметры процесса | | | Тестирование устойчивости к дезактивации | | | |
|---|---|---|---|---|---|---|---|---|---|---|---|---|---|---|
| | | | | | T;°C P; МПа | | 300 3.0 | 340 3.0 | 360 3.0 | 380 5.0 | 300 3.0 | 340 3.0 | 380 5.0 | 300 3.0 |
| | | | | Сырье | | | Содержание серы и азота в ppm; йодное число в г$J_2$/100г | | | | | | | |
| | | | | | | | S-11000, N-0.095; Й.ч.-0.3 | | | | S-15000; N- 0.145; Й.ч.-0.7 | | | |
| №1 | 400 | 1.0:0.1:0 | Дизельной фракцией | 1.48 | 2.47 | 12.4 | 2300 | 1887 | 1297 | 620 | | | | |
| АКНМ-3/5 | 400 | 1,0:0,1:0,1 | | 1.48 | 2.51 | 12.5 | 2274 | 1776 | 1207 | 596 | 2280 | 1800 | 413 | 2283 |
| АКНМ-2/9 | 400 | 1,0:0,1:0,1 | | 1.70 | 4,19 | 17.5 | 2111 | 1544 | 995 | 367 N-0.028 | 2113 N-0.09 | 1605 N-0.080 | 381 N-0.037 | 2120 N-0.092 |
| АКНМ-2/9 | 150 | 1,0:0,1:0,1 | | | | | 2733 | 2005 | 1294 | | | | | |
| АКНМ-2/9 | 150 | 1,0:0,5:0,1 | $H_2S$ | | | | 461 | 334 | 275 | | | | | |
| АКНМ/№16 | 400 | 1,0:0,1:0,5 | Дизельной фракцией | 1.0 | 2.3 | 8.5 | 7838 | 6511 | 4650 | | | | | |
| АКНМ/№17 | 400 | 1,0:0,5:0,5 | | 1.44 | 2.36 | 12.3 | 2203 | 1803 | 1215 | 662 | 2224 | 1820 | 415 | 2235 |
| АКНМ/№19 | 400 | 1,0:0,2:0,5 | | 1.39 | 2,23 | 9.03 | 3216 | 2830 | 1573 | 900 | | | | |
| АКНМ/№18 | 400 | 1,0:0,2:0,5 | | 1.46 | 2.48 | 12.1 | 2367 | 1903 | 1238 | 617 | | | | |
| АКНМ/№18 | 120 | 1,0:0,2:0,5 | | | | | 937 | 640 | 255 | | | | | |
| АКНМ/№18 | 120 | 1,0:0,2:0,5 | $H_2S$ | | | | 530 | 379 | 283 | | | | | |
| АКНМ/№15 | 400 | 1,0:0.2:0.3 | Дизельной фракцией | 1.47 | 2.50 | 12.3 | 2242 | 1682 | 1165 | 562 | | | | |
| АКНМ/№15 | 150 | 1,0:0.2:0.3 | | 1.47 | 2.50 | 12.3 | 2868 | 2150 | 1430 | | | | | |
| АКНМ/№15 | 150 | 1,0:0.2:0.3 | $H_2S$ | 1.47 | 2.50 | 12.3 | 530 | 378 | 286 | | | | | |
| АКНМ/№20 | | 9:0:1 | Дизельной фрак. фрфракцией | 1.49 | 2.49 | 12.2 | 5442 | 4352 | 3123 | 1644 N-0.045 Й.ч.-0.2 | 5835 N-0.120 Й.ч.-0.3 | 4889 N-0.113 Й.ч.-0.2 | 1704 N-0.075 Й.ч.-0.1 | 5987 N-0.125 Й.ч.-0.3 |

synthesis method and close content of active metals had a positive effect on the stability of the catalysts under hydrotreating conditions. Tests of a series of catalysts according to another accelerated method, i.e. on fine grain (1.0-1.5 mm) and high volumetric rate (6.0$h^{-1)}$ at temperature 300°C and pressure 3 MPa revealed that the stability changes in the series: AKNM-3/5> AKM> AKNM-4/11≥ ANM-2/3> AKNM-4/16.

The developed catalyst AKNM-3/5 was long-term tested at the pilot plant under mild conditions of hydrotreating process, close to those used at the industrial unit of oil hydrotreating G-24 of Fergana Oil Refinery (Table 5.4). It was found that during 1000 hours of testing the pressure drop on the catalyst layer did not increase, fluctuations of sulfur content in hydrogenysate of deasphaltized residual feedstock were within 0.50 - 0.52 %, and chromaticity from 5.2 to 5.3 units. CNT. It was experimentally found that the stability of the developed catalyst increases in case of using a protective layer [250; C.40, 251; C.314], at the same time the quality of the hydrogenizate is improved. The main physical and chemical characteristics of the hydrogenizate quite corresponded to the required parameters for the production of base oils.

The technology of production of optimal trimetallic catalyst AKNM-3/5 on industrial equipment according to the proposed scheme (Fig.3.6) has been worked out and pilot technological regulations for production of catalyst AKNM-3/5 for oil hydrotreating on industrial equipment of UzKFITI and Technical Conditions for it have been developed. A pilot batch of the catalyst was produced and loaded into the industrial reactor (No. R-2 II flow) of oil hydrotreating unit G-24 of Fergana Oil Refinery.

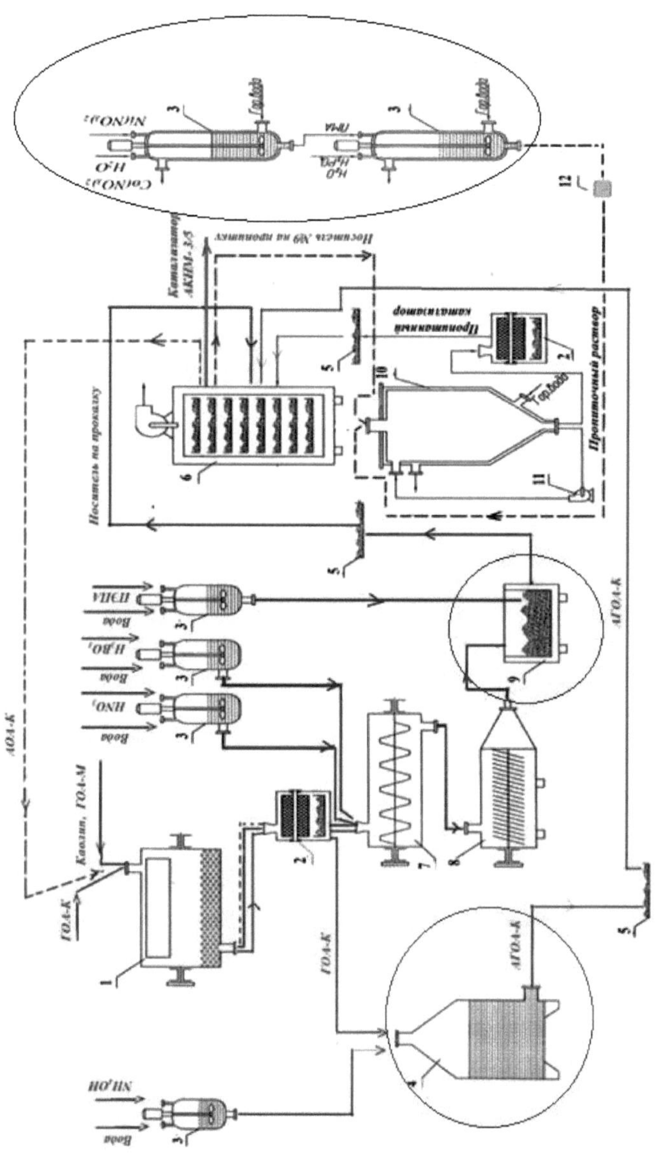

*1-ball mill, 2-sieve, 3-mortar container, 4-cooker, 5-baking pan, 6-milk oven, 7-mixer, 8-extruder, 9-texture modifier container, 10-impregnator, 11-pump, 12-pH-meter.*

**Fig. 3.6. Technological scheme of production of cobaltnickel-molybdenum catalyst for hydrotreatment of oils AKNM 3/5**

# CHAPTER IV. TECHNOLOGIES OF TRIMETALLIC CATALYST SYNTHESIS

## §4.1 Interactions of cobalt, nickel and molybdenum salts with the carrier in the process of catalyst synthesis

### §4.1.1 Synthesis of catalysts

As follows from the above literature review, to obtain high-quality fuels and oils, the corresponding oil fractions are hydrotreated from sulfur and nitrogen compounds on Al-Co-Mo or Al-Ni-Mo catalysts containing 15-18% of active components, the physical and chemical characteristics of which are studied in detail. However, patents have already been registered for the preparation of catalysts containing simultaneously three active components: cobalt, nickel, and molybdenum, combining the positive properties of both types of binary catalysts [183; 45-50, 228; P. 1,].

In particular, to obtain such systems it is proposed to use either spent thermally regenerated aluminocobalt-molybdenum catalysts (ACM) or pellet sludge from ACM catalyst plants, which is used as a source of cobalt and partially molybdenum. Nickel and missing molybdenum are introduced by various methods into the molding mass containing ground ACM and fresh aluminum hydroxide. Structure, phase composition, morphology and size of aggregates of active metals in the obtained catalysts are not given, meanwhile hydrodeazotizing and hydrodesulfurizing activity of the catalysts depend on the amount and ratio of active components, method of their introduction, modifiers used, etc. Deep desulfurization of oil fractions, depending on the ratio of thiophene compounds and alkyl derivatives of dibenzthiophene, requires sequential carrying out of the process in reactors with different partial pressure of hydrogen, as well as the use of both Co-Mo and Ni-Mo catalysts. In the processing of feedstocks with predominance of alkyl sulfides, mercaptans and alkyl polysulfides, the main task of hydrotreating is hydrodesulfurization. According to the recommendation, Lulic P. [229; P. 585], it is best to place a cobalt-molybdenum catalyst in the main bed (or the first reactor along the course). When it is necessary to maximize the hydrogenation of aromatic hydrocarbons and deazotization, it is better to choose nickelmolybdenum catalyst as the main bed. Aleksandrov P.V. and co-authors [230; P.7-8] propose to load nickel-molybdenum catalyst in the tail part of the reactor with a higher temperature favorable for hydrodeazotization to ensure maximum efficiency of the combined bed reactor. Recently, works indicating the expediency of research of catalysts according to the "two-in-one" principle have begun to appear. Catalysts combining the positive properties of cobalt and nickel in the composition of one multicomponent molybdenum-containing catalyst were synthesized by various methods [231; P.56-58, 232; P.139, 233; P.8311].

We have tested the following methods of applying active components to the surface of alumocaolin carriers to obtain trimetallic Co-Ni-Mo catalyst.

1 - impregnation of carriers with half of the calculated amount of aqueous solution of ammonium paramolybdate, intermediate heat treatment at different temperatures and re-impregnation with a joint solution of ammonium paramolybdate and nickel and cobalt nitrates stabilized with phosphoric acid [234; P.126-127].

2 - impregnation of carriers with a calculated amount of aqueous solution of ammonium paramolybdate, intermediate heat treatment at different temperatures and re-impregnation with a joint aqueous solution of nickel and/or cobalt nitrates [217; P.142, 234; P.126, 235; P.111, 236; P.12, 237; P.122]. When obtaining catalysts with the content of $MoO_3$ more than 10 %, phosphoric acid was added to the solution of ammonium paramolybdate.

3 - by single impregnation with a joint solution of ammonium paramolybdate and nickel and cobalt nitrates stabilized with phosphoric and/or citric acid [217; P.142,222; P.23,234; P.126-127, 236; P.12, 238; P. 42, 239; P.48-53, 240; P.104, 241; P.125].

4 - by co-extrusion of a mixture of appropriate aluminum hydroxides, kaolin and a joint solution of ammonium paramolybdate and nickel and cobalt nitrates stabilized with phosphoric acid [218; P.45, 239; P.49].

Taking into account the moderate moisture absorption of synthesized carriers, as well as the reduced solubility in water of ammonium para-molybdate produced by UzKTZhM JSC in Chirchik, phosphoric acid was used in the preparation of concentrated impregnation solutions. It is known that when phosphorus compounds are introduced into the impregnation solution, the latter has a stabilizing effect due to the formation of heteropoly compounds when interacting with molybdate anions. As a result, during impregnation the carrier adsorbs more Ni(Co) and Mo, as a result the activity of the catalyst, primarily deazotizing, increases [66; P.232]. The possibility of formation of heteropoly compounds at certain stages of catalyst production by conventional impregnation also follows from literature sources [118;P.921,133;P.905,188;P.47]. Since the composition of heteropoly compounds strongly depends on the ratio of components, the pH of the solution, and the surface properties of the carrier, we paid special attention to this problem when studying catalyst samples and model systems. Therefore, actively involving for discussion the results of published studies of composition and physicochemical properties of individual heteropoly compounds, we parallel the comparative data we obtained under identical conditions for applied and massive samples. In addition to trimetallic catalysts containing molybdenum, cobalt, nickel, and phosphorus on the optimal carrier No. 9, or carrier No. 8 and some others, model systems excluding one or more of the listed components were synthesized.

The first series of specimens was obtained by double impregnation, with intermediate drying and calcination of the intermediate product for preliminary thermal consolidation of the first applied metal. The order of application was varied. For example, to obtain trimetallic catalyst No. 1-AKNM, first, an aqueous solution of ammonium para-molybdate (sample No. 1 - AM) was applied to the carrier No. 9, then, after intermediate heat treatment, the pellets were impregnated with a mixed aqueous solution of cobalt and nickel nitrates, after which the pellets were subjected to heat treatment at 400 and 550°C. The catalyst (No. 1-AKNM) after calcination at 550°C contains 10.5 % molybdenum oxide, 2 % nickel oxide and 1.5 % cobalt .

To obtain catalysts of the second series No. 2-AKNM-P, with the same content of transition metals as No. 1-AKNM, but with a single impregnation, first a given amount of ammonium para-molybdate was dissolved in water, the pH of the solution was brought to +3 by adding phosphoric acid. Then separately prepared solutions of nickel and cobalt nitrates in water were added and the resulting complex solution was impregnated into the carrier №8. Catalyst No. 2-AKNM-R, obtained by single impregnation, after calcination at 550°C contains 10.5 % of molybdenum oxide, 2 % of nickel oxide, 1.5 % of cobalt oxide and 2.0 % of phosphorus oxide. It is known that single impregnation of the carrier with a solution, two or more, of transition metal salts is used quite widely and gives very good results. However, there are several points to be taken into account during synthesis. These are the problem of mutual solubility of salts in certain conditions, the possibility of sorption of only one component in the absence of sorption of the other, deposition in the pores of the carrier, in the first place, little soluble substance. The first problem was solved by stabilizing the impregnation solution with phosphoric acid [217; P.143-144, 238; P.42]. The choice of stabilizer was conditioned by the fact that the introduction of orthophosphoric acid not only prevents precipitation at the impregnation stage, but also directly affects the surface properties and determines the structure of the applied active phase due to the formation of heteropoly compounds. The role of phosphorus compounds in the formation of active centers of hydrotreating catalysts is not obvious and is still a subject of discussion [118; P.914, 131; P.41-42]. However, most authors consider undesirable the phosphorus concentration (in terms of $P_2O_5$) exceeding more than 5 wt.%. Therefore, for stabilization of concentrated joint solutions of ammonium paramolybdate with nickel and/or cobalt nitrates, in a number of experiments a part of phosphoric acid was replaced by citric acid [222;

P.23. 239; P.51, 240; P.104]. The ciphers of catalysts prepared by impregnation of solutions stabilized by phosphoric acid alone contain the letter "P", and by a mixture of phosphoric and citric acids are designated as "P, Cit".

### §4.1.2 Thermographic study of catalysts

The influence of the stabilizer, phosphoric acid, on the interaction of impregnation solutions with the carrier surface is well demonstrated by differential thermal analysis (DTA) [237; P.122-123]. The DTA curves of dried AM catalysts obtained by impregnation of carriers (No.3, No.8 and No.9) with ammonium paramolybdate solution are characterized by: deep endo effect150°sobtained by removal of adsorbed water and ammonia, and weak, strongly chemisorbed ammonium ions. On the DTA curve (Fig. 4.1) of dried NA and AK catalysts obtained by impregnation of carriers (№3, №8 and №9) with nickel or cobalt nitrate solution, there were only endo-effects at 145-150°C, caused by removal of adsorbed water and products of thermolysis of nitric acid salts and accompanied by weight loss [84; P.162].

If not calcined GOA-M was impregnated with nickel nitrate, i.e. modeled the coextrusion method, there was a clear decrease of endothermic effects from the decomposition of pseudobemite and bemite in the region of 340-500°C, compared to the initial GOA-M, and a small exo-effect appeared around 230°C, attributed on the basis of literature data to the formation of Al-Ni- bonding precursor of surface aluminonickel spinel. If the co-extrusion of GOA-M and nickel nitrate was carried out in the presence of phosphoric acid, the exoeffect at 230°C did not appear.

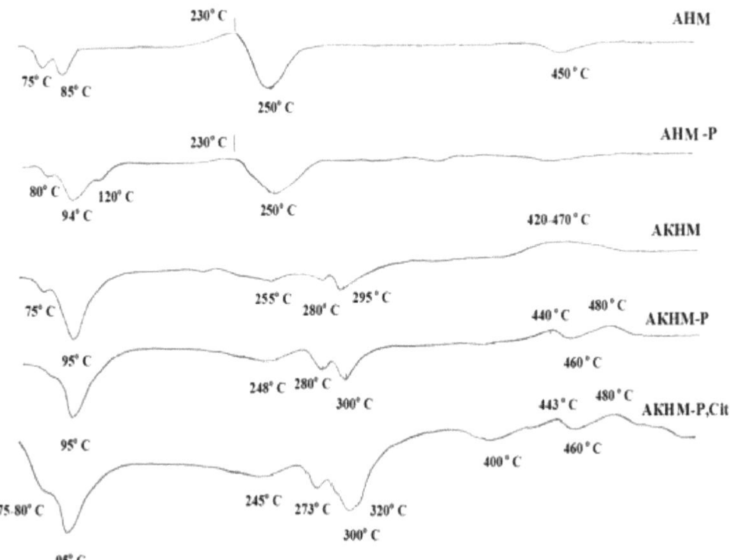


**Figure 4.1. DTA curves of water bath dried catalysts of different compositions.**

From comparison of DTA curves (Fig. 4.1) of water bath-dried catalysts obtained by co-extrusion of GOA-M and kaolin with solutions of nickel nitrate and ammonium paramolybdate (ANM) or their joint solution with phosphoric acid (ANM-P), one can see almost complete disappearance of the exo-effect at 230-250°C in the case of using the joint solution stabilizer.

The DTA curves of NiO (2%)-$MoO_3$(10.5 %) catalysts on carriers #3, #8 and #9 (not given), both at double (ANM) and single impregnation (ANM-P and ANM-P, Cit) showed minor thermal effects at 120, 180 and 250-280°C. These effects are due to the detachment of water and ammonia and the formation of new chemical bonds. These thermal effects are close to the thermogram $(NH_4)_{(4)}[Ni(OH)_{(6)}Mo_6O_{18}]$-$5H_2O$ given in [131;C.107], for the thermal decomposition of an individual compound -ammonium salt of nickelmolybdenum heteropolyacid. On the DTA curve of commercial $(NH_4)_{(4)}[Ni(OH)_{(6)}Mo_6O_{18}]$-$5H_2O$ three weak endothermic peaks at 140, 210 and 300°C were observed, caused by the destruction of the existing bonds in the molecule of heteropolyacid, followed by the release of water and ammonia. According to the data given in the book [84; P.163], the formation of $NiMoO_4$ phase is accompanied by an exothermic effect at 280-300°C.

Some shift of endothermic effects on DTA curves of ANM-P and ANM-P, Cit catalysts to the low-temperature region may be due to the formation of surface Al-Ni-Mo structures. On thermograms of all studied catalysts there were high-temperature thermal effects at 750-860°C due to phase transformations$\gamma$ - $Al_2O_3$. It should be noted that the appearance of DTA curves of air-dry bimetallic and trimetallic catalysts indicates the parallel occurrence of both thermal processes with heat release and its absorption in the same temperature range of 200-300°C. According to [131; P. 110-111] endoeffects at 280-300°C were attributed to the loss of water entering different spheres of the gamut of coordination compounds of molybdenum type: Al - Mo, Ni - Mo, Co -Mo, Al - Ni - Mo, Al - Co - Mo and others. Hydrated Ni -Mo, Co -Mo and Mo -P compounds are formed at the stage of preparation of impregnation solutions, and Al - Ni - Mo, Al - Co - Mo and Al - Co - Mo are the product of their subsequent interaction with Al - OH groups of the carrier. The reactions of hydrated compounds of impregnation solutions with surface hydroxyl groups were even more active at the mixing stage during the preparation of catalysts by coextrusion [218; P. 142-143]. This interaction was manifested by a strong decrease in the characteristic endothermic effects from the decomposition of pseudobemite and bemita in the 340-507°C region, along with the appearance of new exo- and endothermic effects at 277 and 287°C, corresponding to the rearrangement of complex associates. At lower temperature, crystallization water from the outer coordination sphere of heteropoly compounds and hydrated cobalt and nickel molybdates was generally removed. At temperatures close to 300$^{(o)o,}$ the so-called "zeolite" water is removed from the inner coordination sphere of heteropoly compounds. The appearance of weak exo effects at 440 - 480°C, more pronounced in the thermograms of NiO-$MoO_3$ (ANM-P and ANM-P, Cit) than of CoO-NiO-$MoO_3$ catalysts (ACNM-P and ACNM-P, Cit) prepared by single impregnation with joint solutions, is characteristic of the bond rearrangement processes in Ni - O - Mo, Co - O - Mo, Al - O - Mo, P - O - Mo associates, which are part of the corresponding heteropolyanions. The low intensity of the peaks is due to the fact that some of the bonds in heteropolyanions of heteropoly compounds are broken, expending energy, and the formation of new bonds is accompanied, on the contrary, by the release of energy [131; P.109]. The 440 - 480°C exo effects are not manifested in the case of CoO-$MoO_3$ catalysts (ACM-P and ACM-P, Cit), which indicates the thermal lability of cobalt heteropolymolybdates, which according to literature data decompose already at heat treatment around 300°C. On thermograms (Fig.4.1) of the samples of catalysts No.1 - AKNM and No.2 -AKNM-P dried on a water bath clearly show low-temperature endo-effects from detachment of adsorbed water and ammonia, as well as crystallization water from hydrated nickel and cobalt monomolybdates, accompanied by weight loss. Exothermic effects in the region of 440-480°C, typical for thermal transformations of heteropoly compounds with the removal of water from the inner coordination sphere, are barely visible in the thermogram of catalyst No. 1-AKNM against the background of the resulting DTA curve from the layering of a series of thermal processes occurring with the absorption and release of heat in this temperature range. The almost imperceptible weight loss in this temperature interval indicates a low content of bound water. In the case of catalysts #2 -AKNM-P and AKNM-P, Cit, the exo effects in the region 440-480°C appear more clearly (Fig.4.1). These peaks on the DTA curves may correspond to the processes of rearrangement of associates accompanied by the breakage of part of the existing Ni-O-Mo, Co-O-Mo, Al-O-Mo, and P-O-Mo bonds, possibly included in the heteropolyanions (energy expenditure), and the formation of new bonds (energy release) [131; 105-106]. The appearance of a very broad exothermic effect, without a pronounced maximum on the DTA curve of the catalyst AKNM-P, Cit, is due to the gradual decomposition of the organic component accompanied by weight loss.

### §4.1.3 Examination of catalysts by spectral methods

Changes in the structure of transition metal compounds in the process of obtaining catalysts by different methods also follow from the electronic diffuse reflectance spectra (ESDO) presented in Figures 4.2. and 4.3. The absorption bands at 26.72; 15.34 $kcm^{-1}$in the ESDO of catalysts AN-P (Fig. 4.2 A) and AN-P, Cit dried on a water bath are characteristic of nickel ions in octahedral coordination and correspond to the electronic transitions $^{3}A_{(2)(g)} \rightarrow (^{3)}T_{(1)(g)}(P)$ and $^{3}A_{(2)(g)} \rightarrow (^{3)}T_{(1)(g)}(F)$[161;P. 1514-1515]. The slight shift of the $Ni^{2+}{}_{Oh}$ bands, relative to the position of the bands in nickel aquacomplexes (Fig. 4.3, B) is due to the incorporation of alumina ligands on the carrier surface into the coordination sphere of nickel ions. In the ESDO of the air-dry NA sample, the absorption bands characteristic of $Ni^{2+}{}_{(Td)}$ ions 15.7 and 16.6 $kcm^{-1}$ prevail (Fig. 4.2 A) described in [112; P. 187, 218; P. 50].

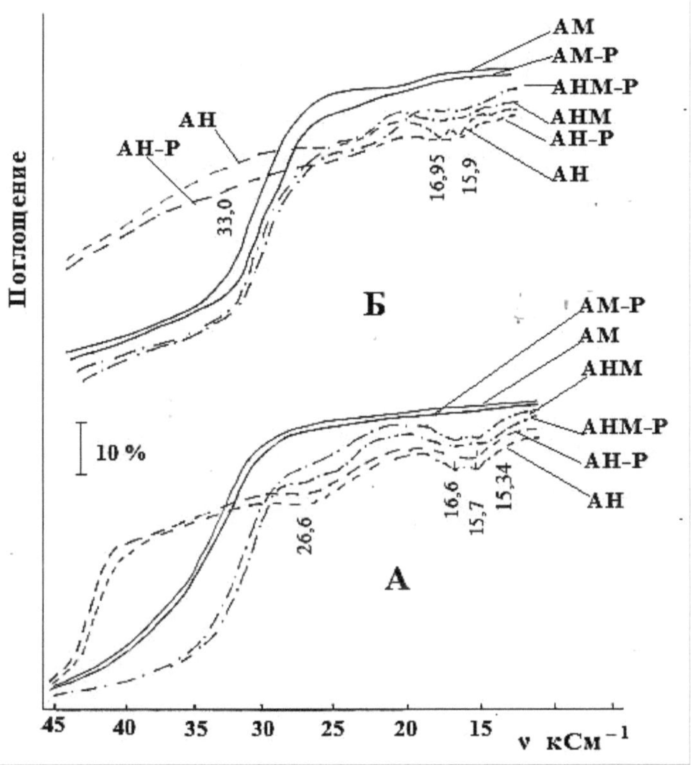


**Fig.4.2. Electronic spectra of diffuse reflection of catalysts on the carrier No.8.**
**A - air-dry, B - calcined at 550°C.**

After calcination at 400° C, and even more so at 550° C, spectra of Al-Ni systems show bands at 16.95 and 15.90 kcm$^{-1}$ characteristic of $Ni^{(2+)}{}_{(Td)}$ions in the $NiAl_2O_4$ spinel structure, but their intensity is less in the spectra of AN-P, Cit and AN-R (Fig. 4.2 B). Thus, surface phosphorus compounds formed on the carrier No. 8 and others prevent the formation of bulk spinel structures, which agrees well with literature data for systems prepared on aluminum oxide in the absence of kaolin [110; P.1267, 112; P.192]

When cobalt nitrate from aqueous (catalyst AK), aqueous-phosphoric acid solution (catalyst AK-P) is applied to the carrier No.9, similar processes occur as well as in the presence of phosphoric and citric acids. Although the compounds of cobalt with the carrier do not appear in the X-ray diffraction patterns of samples with low content of CoO (1.5%), however, these structures are clearly identified in the electronic spectra (Fig. 4.3).

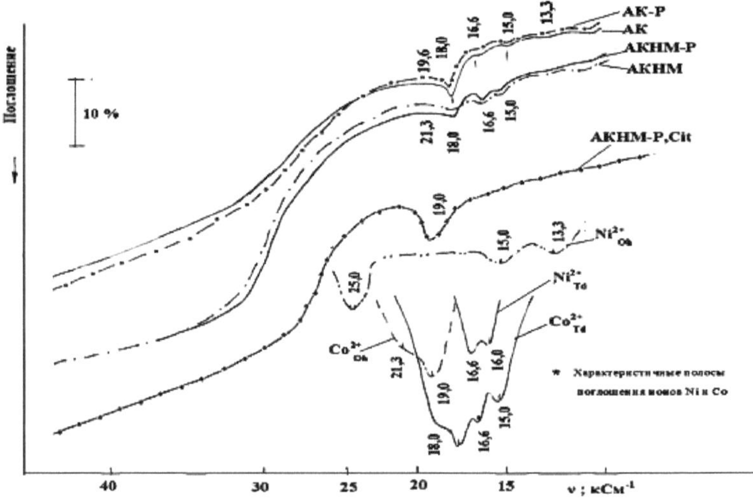


**Figure 4.3. Electron diffuse reflectance spectra of catalysts calcined at 550°C**

**AKNM-P, Cit catalyst dried at 120°C.**

Already in the process of impregnation and drying, the bands 19.5 and 21.3 $kcm^{-1}$ specific for octahedrally coordinated $Co^{(2+})$ ions in the initial salt $Co(NO_{3})_{(2)}$·$6H_{(2)}O$ almost completely disappear. Instead, a new system of absorption bands appears. A band at about 6.0 $kcm^{-1}$ and a triplet with maxima at 15.0, 16.6, and 18.0 $kcm^{-1}$ from the $Co^{(2+})_{(Td)}$ ions [236; P.12-13]. The weak band at 10.3 $kcm^{-1}$ may belong to both $Co^{(2+})_{(Td)}$) and $Co^{(3+})_{(Oh)}$ ions. Due to thermal instability, adsorbed aquacomplexes of $Co^{2+}_{Td}$ ions decompose, preserving the tetrahedral coordination, regardless of the presence of $H_3RO_4$.

In the ESDO of samples obtained by impregnation of alumina kaolin carriers with acid-stabilized ammonium paramolybdate solutions, the absorption level in the region of 33.0 $kcm^{-1}$ is increased, compared to impregnated aqueous solutions (Fig. 4.2.A). Such a shift in the edge of the $Mo^{6+}$charge transfer band indicates a greater degree of polymerization of $MoO_4^{2-}$ ions anchored on the surface of AM-P,Cit and AM-P compared to AM obtained in the absence of acid stabilizers of the impregnation solution. On the carrier No. 19 the presence of $[Si(Mo_{12}O_{40}]$ is fixed and at calcination of samples of molybdenum-containing catalysts a mixture of isolated molybdenum ions and polymolybdate ions is spectrally manifested, the first structures being rather firmly fixed on the carrier surface by Al-O-Mo bonds.

Accordingly, in the Raman spectra (CR) of the AM sample, a broad band at about 915 $cm^{-1}$ characteristic for $MoO_4^{2-}$ ions prevailed[84; P.142]. In the Raman spectrum of AM-P, on the contrary, the band at 950 $cm^{-1}$ characteristic for

$Mo_7O_{24}^{6-}$ ions was predominant, and the 915cm$^{-1}$ band appeared as a shoulder in its background.

This conclusion is confirmed by the results of selective extraction of molybdenum compounds from a series of catalysts according to the procedure given in [112; P. 186-190]. At close content of molybdenum (≈ 12 wt. % MoO(3)) and the same carrier. $MoO_3$) and the same carrier, the amount of water-extractable (i.e. weakly bound to the carrier) molybdenum is slightly higher in the case of catalyst №2-AM-P/#9 (9.65% wt. % $MoO_3$). $MoO_3$), comparison with #1-AM/#9 (6.43% wt. $MoO_3$). It increases to 7.13 wt. % MoO(3)), with 7.13 wt. % MoO(3)). $MoO_3$), when transferred to a similar sample of AM-P, Cit/#9, but obtained by single impregnation with a solution stabilized with a mixture of phosphoric and citric acids. ESDO in the ultraviolet spectral region 29.4-33.3 kcm$^{-1}$ (Fig. 4.2) of AM and AM-P catalysts only confirm the known data on the predominance of octahedrally coordinated molybdenum in the studied samples, but do not allow to distinguish octahedrally coordinated molybdenum in isopoly compounds from $Mo^{6+}{}_{Oh}$ in molybdenum oxide. To prove the formation of isopolymolybdates in the catalysts studied by us, UV absorption spectra of aqueous and ammonia extracts (Fig.4.4) obtained in the process of selective extraction of molybdenum compounds from AM and AM-P catalysts were taken. The presence of absorption bands at 48.0 and 43.5 kcm$^{-1}$in the spectra of aqueous and ammonia extracts from both samples proved the presence of tetrahedrally coordinated molybdenum in the catalysts.

The manifestation of more intense band 41.0xm$^{-1}$ in the spectrum of AM-P aqueous extract, in comparison with AM, according to literature sources, indicated a higher concentration of molybdenum isopolyanions when the carriers were impregnated with a solution with low pH value.

From a series of VNIINP works devoted to the study of hydrodessulfurization catalysts and systematized in Chukin G.D.'s book. [84; P.163-173], it is known that during the preparation of aluminonickelmolybdenum catalysts by co-precipitation surface structures are formed, which can exist in the form of nickelmolybdenum associates, including $OH^-$ groups, ammonium cations and nitrate anions, and these associates are characterized by weak interaction with the carrier. Conducting similar experiments, i.e. combining thermography, electron and CR spectroscopy with the method of selective extraction of $Mo^{6+}_{Oh}$, $Mo^{(6+)}_{(Td)}$, $Ni^{(2+)}_{Oh}$, $Ni^{(2+)}_{(Td)}$, $Co^{(2+)}_{(Oh)}$ and $Co^{(2+)}_{(Td)}$, as applied to the catalysts synthesized by us, allowed us to prove the formation of structures where molybdenum ions are associated with transition metal cations and/or surface aluminum cations. Since the same series of absorption bands was observed in the spectrum of the water-bath dried catalyst №1-ANM, obtained by successive impregnation of the carrier № 9 with aqueous solution of ammonium molybdate and nickel nitrate with intermediate thermal fixation of molybdenum ions, it is possible to assume the formation of similar structures in the studied system. The formation of similar surface mixed oxide complexes can be judged by the shift of the $Ni^{2+}_{Oh}$band in the aluminonickel system from 26.8 $kcm^{-1}$to 24.5 $kcm^{-1}$ in the catalyst No. 1 -ANM, which is due to the introduction of $[MoO_4]^{2-}$ anions into the coordination sphere of $Ni^{(2+)}$ ions. The parameters of the spectrum are close to those of the spectrum of $NiMoO_4$ with characteristic bands at 24.0 and

13.0 kcm$^{-1}$ (Fig.4.2). According to the results on selective extraction of nickel and molybdenum, when using for synthesis the carrier No.9 containing AOA-K and treated with texture modifier, the amount of Mo and Ni structures extracted from the catalyst by aqueous-ammonia solution increases to 2.2 % $MoO_3$ and 0.03 %NiO from the initial one. For catalysts on carriers #1-#4 (Table 3.2), this value was only 0.6 % $MoO_3$, and nickel ions were not detected in the aqueous-ammonia extracts. That is, an increase in the amount of compounds such as surface aluminum molybdate and nickel aluminate - the most active in breaking C-S bonds - was observed.

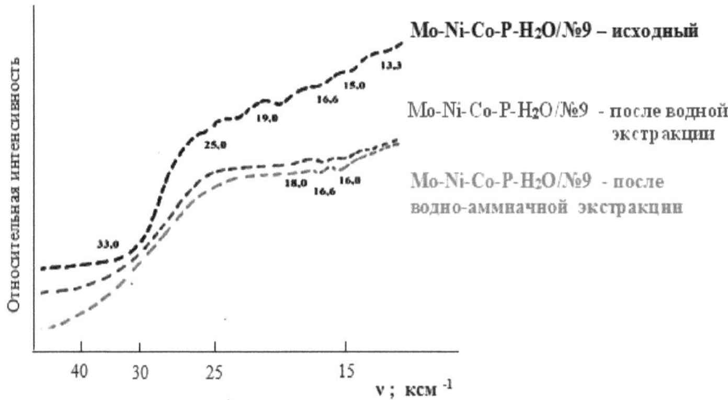


**Figure 4.5. Manifestation of removal of structures with transition metal ions from catalysts in the process of selective extraction (electron diffuse reflectance spectra).**

In the electronic spectra of catalyst №2-ANM-P, obtained by one-stage impregnation of the carrier №8 with a complex solution of nickel and molybdenum salts stabilized by phosphoric acid, the intensity of bands from octahedrally coordinated ions of nickel, at 27.0 -24.0 and 15.4-13.0 $kcm^{-1}$, and molybdenum in the region of 35.0-32.0 $kcm^{-1}$, in comparison with the catalyst №1-ANM, prepared by double impregnation without phosphoric acid (Fig. 4.2).That is, in the presence of phosphoric acid there was a deepening of the interaction between aquacomplexes of nickel and molybdenum with the formation of surface hydroxylated aluminonickel-molybdenum associates [218;P.50]. At the same time, we cannot completely exclude the possibility of competing interaction of phosphoric acid with the carrier, when poorly soluble aluminum phosphates will screen the surface centers of the carrier, preventing the interaction of nickel and molybdenum compounds with hydroxyl groups of aluminum oxide, thus creating favorable conditions for their reaction with each other.

We found that the spectra of trimetallic catalysts on all studied carriers were characterized by an abundance of absorption bands and inflection points. A series of bands 27-23, 21.0, 19.2 and an inflection point around 13.0 $kcm^{-1}$ in the electronic spectra indicated the formation of cobalt and nickel molybdates. The high extinction of triplet $Co_{Td}^{2+}$ bands at 15.0, 16.6 and 18.0 $kcm^{-1}$ overlapping the absorption bands characteristic of $Ni_{(Td}^{)(2+)}$ ions (16.8 and 15.8 $kcm^{-1}$), $Co_{oh}^{(2+}$)(19.0 $kcm^{-1}$) and $Ni_{(o)(h)}^{2+}$( 15.0 $kcm^{-1}$), did not allow an unambiguous assessment of the ratio of octahedrally and tetrahedrally coordinated ions. But the disappearance of the absorption band of $Mo^{(6+}{}_{)(Oh)}$ at 41.0 $kcm^{-1}$, characteristic of weakly bound to the surface of nickel isopolymolybdates and aluminomolybdates, while the intensity of the bands 48.0 and 43.5 $kcm^{-1}$ in the spectra of the catalysts, after selective extraction with water testified to the predominance of tetrahedrally coordinated molybdenum ions. The study of the state of active components, both in the initial catalysts and after the separation of weakly and strongly bound to the surface of the applied metal compounds, showed that after water and, to an even greater extent, water-ammonia extraction, the intensity of absorption bands 34.0-33.3 $kcm^{-1}$, which could be attributed to surface polymolybdate structures, decreased in the ESDO.

This applies to a lesser extent to nickel ions in octahedral coordination in the structure of nickel molybdates and nickel-molybdenum asso-ciates (poorly resolved bands in the region 23.0-27.0 and 14.0-12.0 $kcm^{-1}$). The spectrum with a broad unresolved band in the region of 18.3-15.8 $kcm^{-1}$, which is probably a superposition of bands from $Ni^{(2+})_{(Td})$ and $Co^{(2+})_{(Td})$ ions, slightly changes, the absorption in this region slightly increases, and the bands corresponding to nickelmolybdate associates, on the contrary, decrease. It is characteristic that at selective extraction with water the degree of extraction of cobalt from trimetallic catalysts on carriers No. 3 and No. 5 was much higher than that of nickel. In the case of catalysts on carriers No. 8 and No. 9, the opposite picture was observed. The presence of kinks in the regions 27-23 and 17-13 $kcm^{-1}$ did not contradict, but also did not prove, the formation of -Al-O- Ni-O-Mo- and -Al-O-Co-O-Mo- associates [218; P.50]. Upon alkaline hydrolysis in ammonia solution, both nickel and cobalt were extracted in small amounts, in contrast to molybdenum (Tables 4.1, 4.2) [234; P.127-128]. After completion of the ammonia-water extraction, kinks in the 27-23 and 17-13 $kcm^{-1}$ region of the electronic spectra were no longer observed, due to the removal of associated structures from the surface of the catalysts.

A clear increase in the amount of molybdenum, aluminum, and phosphorus extracted with aqueous-ammonia solution at a single impregnation method, compared to the other variants, indicates the active participation of hydroxyl groups of the carrier in the formation of mixed associates sufficiently firmly bound to the surface of the carriers (Tables 4.2) [239; P.51].

**Таблица 4.1**

**Влияние способа синтеза на силу связи активных компонентов с поверхностью катализаторов, по результатам селективной экстракции соединений Mo, Co и Ni**

| Способ | Состав пропиточных растворов (нитрат никеля – Ni; нитрат кобальта – Co; пара-молибдат аммония – Mo; фосфорная кислота – P; лимонная кислота – cit) при различных способах нанесения активных металлов | Температура про-калки; °C | Количество элементов последовательно экстрагированных $H_2O$ и $NH_4OH$; % | | | | | |
|---|---|---|---|---|---|---|---|---|
| | | | Водой (слабо связанные) | | | Водным раствором аммиака (прочно связанные) | | |
| | | | Mo | Ni | Co | Mo | Ni | Co |
| $MoO_3$ (12-12.5%) CoO(2.5-2.6%) NiO(1.4-1.5%) | | | | | | | | |
| Соэкс-трузия и | 0.5 Mo-$H_2O$ и Mo-Ni-Co-P-$H_2O$/№9 | 200/550 | 3.82 | 4.65 | 7.56 | 8.55 | 0 | 0 |
| | 0.5 Mo-$H_2O$ и Mo-Ni-Co-P-$H_2O$ /№9 | 110/550 | 2.05 | 4.08 | 6.45 | 8.87 | 0.02 | 0.04 |
| | 0.5 Mo-$H_2O$ и Mo-Ni-Co-P-$H_2O$/№7 АКНМ-4/11) | 550/550 | 1.89 | 5.03 | 7.48 | 10.2 | 0 | 0 |
| Соэк стру- | ГОА-М и Mo-Ni-Co-P-$H_2O$ | 550 | 6.4 | 3.86 | 8.64 | 11.7 | 0.26 | 0.26 |
| | ГОА-К и Mo-Ni-Co-P-$H_2O$ | 550 | 10.8 | 7.11 | 18.1 | 74.5 | 1.23 | 0.13 |
| $MoO_3$ (9-10 %) CoO(2.5-2.6%) NiO(1.4-1.5%) | | | | | | | | |
| Двукратная пропитка | Mo-$H_2O$ и Ni-Co-$H_2O$/№9 | 550/550 | 1.77 | 1.82 | 5.33 | 2.56 | 0.01 | 0 |
| | Mo-$H_2O$ и Ni-Co-$H_2O$/№7 | 550/550 | 1.65 | 1.93 | 4,40 | 3,05 | 0,03 | 0.04 |
| | Mo-$H_2O$ и Ni- $H_2O$/№9 (АНМ) | 550/550 | 1.82 | 2,04 | - | 2.17 | 0.03 | - |
| | Mo-$H_2O$ и Co-$H_2O$/№9 (АКМ) | 550/550 | 1.91 | - | 3,56 | 2.88 | - | 0.09 |
| | Mo-$H_2O$-P и Co-Ni- $H_2O$/№9 | 400/550 | 4.13 | 1.78 | 8.22 | 10.8 | 0 | 0.03 |

**Таблица 4.2**

**Влияние носителя на силу связи активных компонентов с поверхностью катализаторов: $MoO_3$ (12,0-12,5%) CoO(2,5-2,6%) NiO(1,4-1,5%), полученных однократной пропиткой, по результатам селективной экстракции соединений Mo, Co и Ni**

| Состав пропиточных растворов (нитрат никеля – Ni; нитрат кобальта – Co; пара-молибдат аммония – Mo; фосфорная кислота – P; лимонная кислота – cit) при различных способах нанесения активных металлов | Температура прокалки; °C | Количество элементов последовательно экстрагированных $H_2O$ и $NH_4OH$; % | | | | | |
|---|---|---|---|---|---|---|---|
| | | Водой (слабо связанные) | | | Водным раствором аммиака (прочно связанные) | | |
| | | Mo | Ni | Co | Mo | Ni | Co |
| Mo-Ni-Co-P-$H_2O$/№3 АКНМ 4/16 | 550 | 10,9 | 7,34 | 18,0 | 37,8 | 0,23 | 0,72 |
| Mo-Ni-Co-P-$H_2O$/№5 (АКНМ) | 550 | 12,8 | 10,2 | 21,4 | 33,4 | 0,32 | 0,75 |
| Mo-Ni-Co-P-$H_2O$/№8 (АКНМ 4/11) | 550 | 14,6 | 40,2 | 23,7 | 29,2 | 0,07 | 0,63 |
| Mo-Ni-Co-P-$H_2O$/№9 (АКНМ 3/5) | 550 | 19,3 | 50,3 | 30,6 | 25,6 | 0,03 | 0,72 |
| Mo-Ni-Co-P-$H_2O$/№9 (АКНМ 4/16) | 400 | 20,5 | 52,4 | 34,4 | 25,8 | 0,02 | 0,60 |
| Mo-Ni-Co-P-$H_2O$/№9 (АКНМ-3/5) | 150 | 23,8 | 52,6 | 34,9 | 26,0 | 0.06 | 0,65 |
| Mo-Ni-Co-P-cit -$H_2O$/№9 (АКНМ-3/5) | 400 | 24,3 | 53,8 | 35,2 | 26,3 | 0,11 | 0,17 |
| Mo-Ni-Co-P-cit -$H_2O$ /№9 (АКНМ-3/5) | 150 | 26,8 | 54,2 | 36,9 | 26,7 | 0,15 | 0,23 |
| Mo-Ni-Co-P -$H_2O$/ № 15 | 400 | 40,6 | 62,3 | 68,2 | 14,9 | 12,9 | 1,20 |
| Mo-Ni-Co-P-cit-$H_2O$/№9; (АКНМ-2/9) $MoO_3$ (17.5%) CoO(4,19%) NiO(1.7%) | 400 | 10,9 | 7,34 | 18,0 | 37,8 | 0,23 | 0,72 |

The obtained results on the state of transition metals in the investigated catalysts, on the example of the Mo-Ni-Ci-Co-P-$H_2O$/#9 sample, were confirmed by the CR spectroscopy data (Fig. 4.6)

A series of partially overlapping bands were observed in the spectrum of the initial sample: the most intense bands corresponding to nickel and cobalt ions in the structure of normal molybdates (965, 955, 910, 834, 400 $cm^{-1)}$, weaker ones from the $Mo_7O_{24}^{(6-)}$ polyanion (950, 570, 360 and 220 $cm^{-1}$) and hardly noticeable from the $MoO_4^{2-}$ anion (910 and 320 cm($^{-1)}$) [84; P$^{.142}$]. [84; C.142].

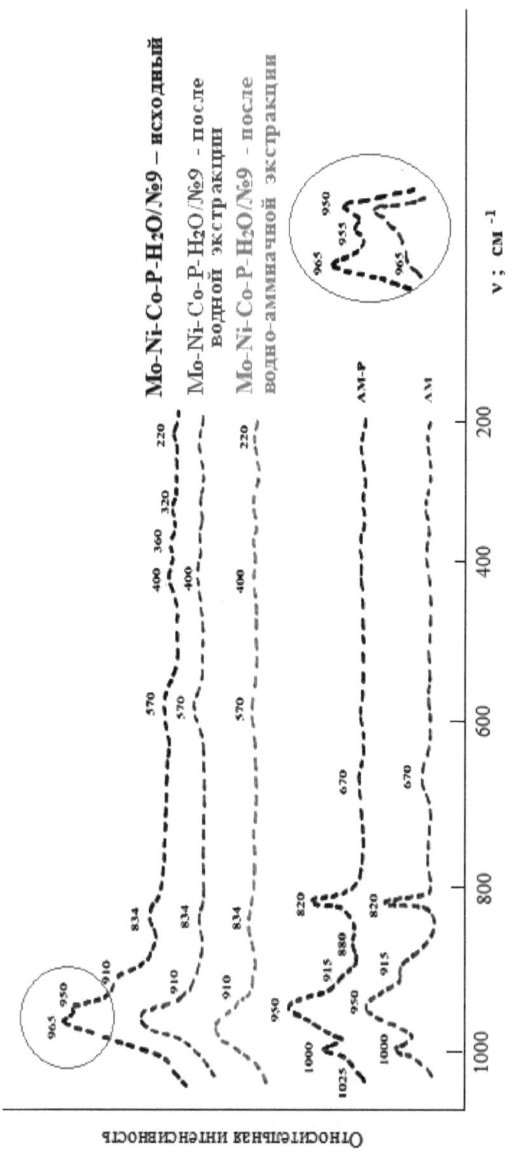


**Figure 4.6. Manifestation of removal of structures with transition metal ions from catalysts in the process of selective extraction (Raman spectra).**

After aqueous extraction of weakly surface-bound bulk molybdates, the intensity of the bands of the first series decreased especially. Water-ammonia extraction decreased to a greater extent the intensity of absorption bands characteristic of molybdenum polyanions strongly bound to the carrier surface.

In the IR spectra, the position of a set of absorption bands in the region 950-880 $cm^{-1}$ corresponding to cis-$MoO_2$-bonds, along with absorption bands in the region 650-450 $cm^{-1}$ attributed in [113; P.154] to bridging M-O-M bonds (Fig.4.6) also unambiguously indicated the presence of molybdates in the composition of catalysts, including polymerized ones

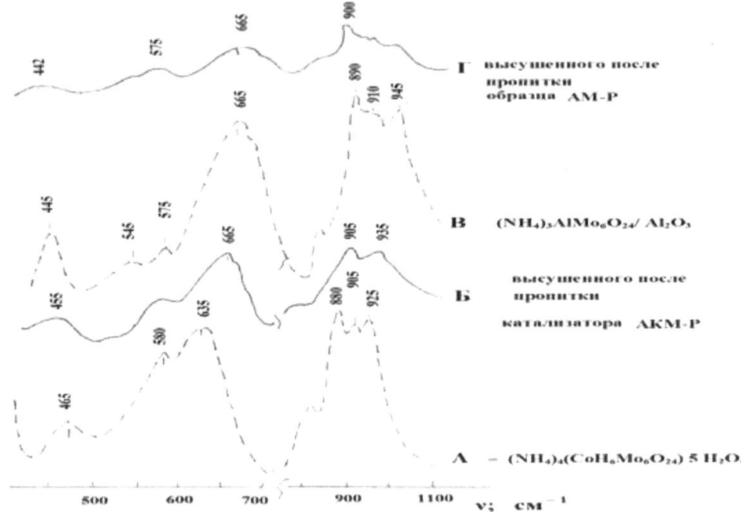


**Figure 4.6. IR spectra of dried AM-P and AKM-P catalysts and heteropoly compounds** (spectra A and B-literature data).

From the comparison of IR spectra of dried AM-P, AKM-P (Fig. 4.6), ANM and AKNM catalysts with literature data for some individual and $Al_2O_3$ heteropoly compounds, it can be concluded that the formation of molybdenum heteropoly compounds with aluminum cation as a central atom is very likely. Since the IR spectra of the catalysts clearly visible absorption bands characteristic of the product of the interaction of ammonium paramolybdate with hydroxyl groups associated surface aluminum cations at low pH values - $(NH_4)_3AlMo_6O_{24}/Al_2O_{(3)}$[113; P.153]

As it was established by spectral studies of heteropoly compounds of the 6th row with Anderson structure, the position of the bands in the IR spectra shifts specifically depending on the central atom (Co, Al, Ni, etc.) in the structure of the molecule. Comparison of the obtained IR spectra of dried catalysts with the IR spectra given in the literature for some heteropoly compounds and catalysts based on them revealed only very weak bands that could be attributed to cobalt or nickel heteropolymolybdates

### §4.1.4 Examination of catalysts by X-ray phase analysis

Diffractograms of trimetallic ACNM (Fig. 4.7), as well as bimetallic ACM and ANM, catalysts have clearly defined reflexes corresponding to the lattice parameters of the carrier components: $\gamma$-$Al_2O_3$ and impurity quartz in kaolin. The resulting structures with active metal ions are X-ray amorphous and appear in the form of a non-symmetric halo

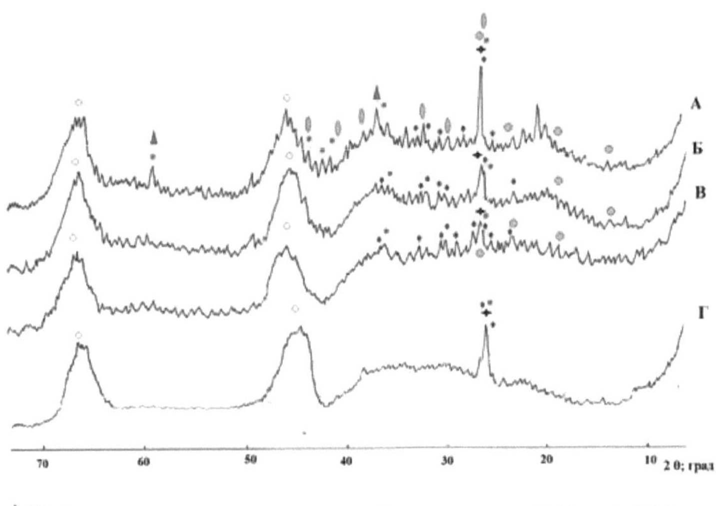


*Impregnation with stabilized solutions;*
*A-ammonia by single impregnation (pH 10) with phosphoric acid:*
*B-single impregnation (pH 2-3),*
*B-double impregnation (pH 2-3);*
*D - with a mixture of citric and phosphoric acids (pH 2-3).*

**Fig.4.7. Diffractograms of calcined at 400°C AKNM catalysts**

Comparatively wide gales of the region of interplanar distances 5.2 - 2.2 Å may indicate the presence of surface compounds with ions of molybdenum, nickel, cobalt, aluminum and phosphorus, but it is not possible to specify these phases and obtain information on their ratio. Volumetric spinel-type phases are not unambiguously X-ray diffracted, although the shape of the diffraction maxima may well indicate their presence, which is not clearly observed due to the overlap of peaks. The smoothed halo contour on the diffractogram of the AKNM-P catalyst indicates a highly dispersed state of molybdenum, cobalt, and nickel compounds, as compared to AKNM-P catalysts also prepared by single impregnation. Analysis of diffractograms of catalysts prepared by single impregnation with a solution of active components stabilized with phosphoric acid or a mixture of acids, calcined at different temperatures, showed that up to a temperature of 400°C the $MoO_3$ phase was absent or in a highly dispersed state. This conclusion also followed from the analysis of the CR spectra, where the characteristic band 820 $cm^{-1}$ was not manifested. The dispersion of bimetallic $NiMoO_4$ and $CoMoO_4$ precursors is also practically completed in this temperature range. The calcination at a higher temperature led to the formation of the crystal phase of excessive $MoO_3$ compared to the stoichiometry of cobalt and nickel molybdates (Table 4.3).

## §4.2 Effect of the method of introduction of active components on the structure of active centers

In order to understand the essence of the processes occurring during the synthesis of complex catalysts, we will dwell in detail on the results of studies of similar model systems obtained by evaporation of impregnation solutions. From literature sources [114; P.456, 133; P. 908] it is known that at the stage of preparation of impregnation solutions molybdenum compounds are most often in the form of polyanions. The degree of polymerization and the probability of formation of heteropolymolybdates with transition metals of group VIII in the impregnation solution depends on many factors. The same applies to changes in the structure of molybdenum poly compounds during interaction with various carriers and their genesis during heat treatment. It is known that the general principle of synthesis of individual heteropolyacids is the acidification of aqueous solutions of monomeric oxoanions to a certain pH value, followed by the separation of polyanions in the form of salts from acidified stoichiometric mixtures of components by various methods. Of great importance is the concentration and order of addition of reagents, the change of which leads to the formation of compounds with different solubility and is accompanied by precipitation, inadmissible in the impregnation of carriers.

In order to reveal the possibility of formation of heteropoly compounds in the process of catalyst synthesis, X-ray diffractograms of model systems, evaporated and dried at 100-150°C solutions used for impregnation of carriers and containing the same concentrations of substances were obtained (Fig.4.8). As can be seen from Figure 4.8 on the diffractogram of the evaporated solution of ammonium paramolybdate with the addition of phosphoric acid crystalline phases do not appear [243; P.194]. The shape of two pronounced haloes with maxima in the region of 6.56 and 3.16 Å suggests the formation of a complex mixture consisting of ammonium polymolybdates $(NH_4)_2OMo_{22}O_{66}$, $(NH_4)_2OMo_{14}O_{42}$, $(NH_{(4))(6)}Mo_7O_{24}$ molybdic acid $NH_2MoO_4$ and some ammonium salt of phosphoromolybdenum heteropolybdate.

In the diffractogram of the evaporated complex solution of nickel nitrate and ammonium molybdate for the preparation of ANM stabilized by phosphoric acid catalyst, the crystal phases of hydrated normal nickel molybdates $NiMoO_4$-$xH_{(2)}O$ and x-$NiOMoO_{(3)}$-$yH_{(2)}O$ are mainly well manifested. Reflexes corresponding to the phosphoromolybdenum heteropolyacid do not appear.

Reflexes from $(NH_4)_3PO_{(4)}$-12$MoO_{(3)}$-3$H_2O$ and $(NH_{(4)})_3PO_{(4)}$-12$MoO_{(3)}$4$H_2O$ - heteropoly compounds of the 12th row were only evident when the concentration of phosphoric acid in the joint solution was increased above 10%. However, the lines corresponding to the interplanar distances of the ammonium salt of heteropolymolybdatanickel - $(NH_4)_4[Ni(OH)_{(6)}Mo_6O_{18}]$-5$H_2O$, which in a number of works is specially introduced into the composition of hydrotreating catalysts to increase activity [135;P.254, 136;P.621], were observed. The formation of heteropolyacids in this system was confirmed by their extraction with ether after preliminary acidification of samples with ammonia removal. During the heat treatment of the evaporated solution for the ANM catalyst, the compound $(NH_4)_4[Ni(OH)_{(6)}Mo_6O_{18}]$-5$H_2O$ was observed by us radiographically only up to a temperature of 350°C (Table 4.3 in the appendix.). After final calcination of the catalyst system at 550°C, anhydrous nickel molybdate was evident in the diffractograms (Figure 4.9B) as a series of moderate intensity reflections with d = 3.68; 3.41; 3.08 ;2.73; 2.32; 1.90; 1.707; 1.629; 1.595; 1.504; 1.41 and 1.394 Å. The bands from $NiMoO_4$ were partially overlapped by strong lines from the hydrated form $NiMoO_4$ $xH_2O$. No crystalline phases of molybdenum oxides were observed in this system. However, rather intense unidentified reflections withd= 6.94; 6.56; 3.48; 1.845 and 1.441 Å are present in the X-ray diffraction patterns.

A mixture of hydrated cobalt molybdates of stoichiometric $CoMoO_4$ - $xH_2O$ and non-stoichiometric $Co_{1-2}$ $MoO_{4-2}$ -13$H_{(2)}O$ composition prevailed in the composition of the evaporated and dried impregnation solution for the preparation of ACM catalysts (Fig. 4.8 B ). The weakly occrystallized phases of unreacted cobalt nitrates $Co(NO_3)_2$ - $H_{(2)}O$ were quite distinct, $Co(NO_3)_{2))(2}$ - $4H_2O$, as well as individual broadened reflections indicating the presence of $(NH_{(4)})_{(4)}CoMo_6H_{(6)}O_{(24)}$-$5H_{(2)}O$, $(NH_4)_6Mo_7O_{24}$ and $MoO_{2-5}(OH)_{(0-5)}$. The non-thermostable ammonium salt of cobalt heteropolymolybdate showed up in the diffractograms only when heat treated to 200°C. After 400°C calcination, a mixture of hydrated and anhydrous cobalt molybdate phases was recorded radiographically (Figure 4.9 B)

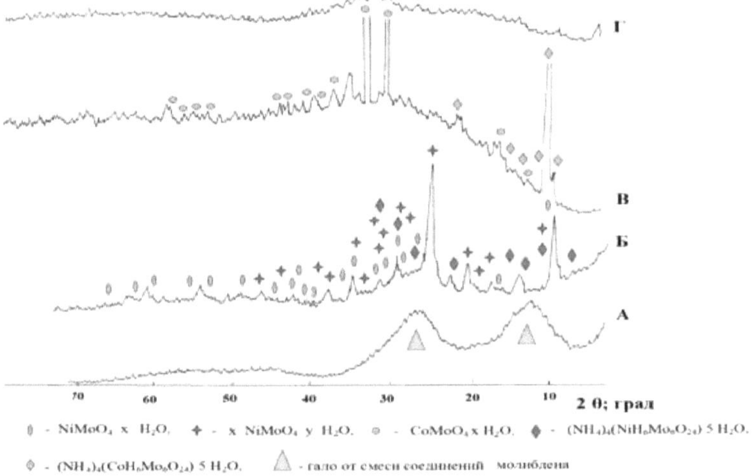


*A - ammonium paramolybdate and phosphoric acid,*
*B - ammonium paramolybdate, phosphoric acid and nickel nitrate,*
*B- ammonium paramolybdate, phosphoric acid and cobalt nitrate,*
*G- ammonium paramolybdate, phosphoric acid, nickel nitrate and cobalt nitrate.*

**Fig. 4.8 Diffractograms of dried on the water bath of model systems.**

As the calcination temperature increased up to 550° C, the intensity of reflections from anhydrous cobalt molybdates increased and a rather well oxidized phase$MoO_3$ appeared.

The diffractogram of the dried impregnation solution for the preparation of trimetallic Co-Ni-Mo catalyst (Fig. 4.8 d) indicated the formation of non-ocrystallized phases. The pronounced halo with a maximum around 3.13 Å can be attributed both to the manifestation of cobalt molybdates ($CoMoO_{(4)}$)-x $H_2O$, $Co_{1\text{-}3}Mo_{4\text{-}2}$ $O_{(4)}$)-x 1.3$H_2O$ $CoMoO_{(4)}$)-0.9 $H_2O$), and to ammonium salts of phosphoromolybdenum, nickelmolybdenum or cobaltmolybdenum heteropolyacids.

The formation of crystalline products occurred only during heat treatment. In the X-ray diagram (Fig. 4.9.D), the most intense bands, we attributed to $CoMoO_4$ and $CoMoO_{(4)}$)-x $H_{(2)}$O and observed in the binary system at 3.38 and 3.28 Å, shift to 3.47 and 3.25 Å, possibly due to the superposition of bands from hydrated nickel molybdate at 3.48 and 3.26 Å. However, the change in the ratio of intensities of characteristic reflections for individual nickel and cobalt molybdates, as well as the disappearance of a number of less intense bands from $CoMoO_4$, although cobalt in this system is three times more than nickel, indicates the formation of mixed cobalt and nickel molybdates. The conducted experiments give grounds for assuming the formation of mixed structures also in the synthesis of trimetallic catalyst on the carrier. The obtained results of thermography, ICS and X-ray phase analysis are in good agreement with literature data that the interaction of heteropoly compounds during impregnation of aluminum oxide increases the thermal stability of the deposited nickel heteropolymolybdate. Comparison of diffractograms of catalysts obtained by single and double impregnation, taking into account the considered literature sources and model systems (evaporated and heat-treated impregnation solutions) allows us to assert that in the case of single-stage impregnation, especially in the presence of citric acid in the impregnation solution, the dispersion of active components in the oxide form is higher (Fig. 4.7).

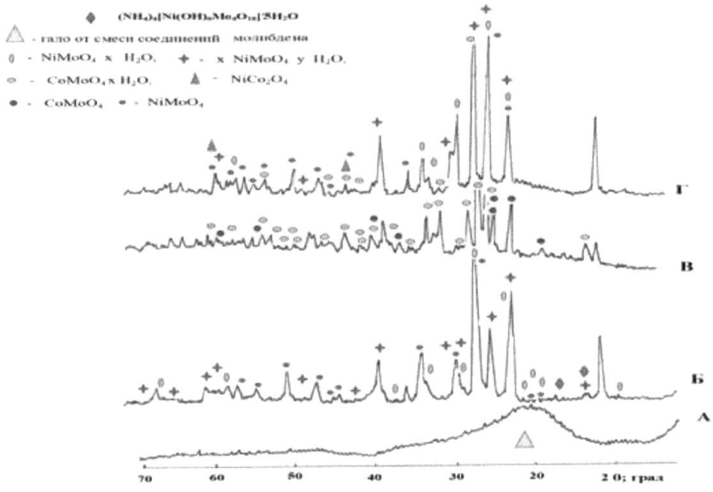


*A - ammonium paramolybdate and phosphoric acid,*
*B - ammonium paramolybdate, phosphoric acid and nickel nitrate,*
*B - ammonium paramolybdate, phosphoric acid and cobalt nitrate,*
*D - ammonium paramolybdate, phosphoric acid, nickel nitrate and cobalt nitrate.*

**Fig. 4.9. Diffractograms of model systems calcined at 350° C.**

According to the results of selective extraction of active components at single impregnation, the maximum formation of structures including Mo, Co and Ni weakly bound to the carrier - oxide precursors of sulfide phases active in hydrogenation and desulfination reactions - was observed. The number of molybdenum structures firmly bound to the carrier, most active in C-S bond breaking reactions also increased, and the number of transition metals not removed by extraction, on the contrary, sharply decreased (Table 4.2).

Comparison of the methods of catalyst synthesis, in terms of the formation of potentially active oxide precursors, revealed the preference for a single impregnation with a joint solution of active components. Therefore, samples for further study of catalytic properties were prepared by this method. Based on the totality of the results of the study of physical and chemical characteristics of carriers and catalysts synthesized by different methods, we have developed a technological scheme for the production of catalysts by a single impregnation method (Fig. 4.10). Table 4.4 presents the results of texture study of catalysts synthesized on carriers obtained on the factory equipment of UzKFITI SEP (Table 3.3). From comparison of Tables 3.3 and 4.4 it follows that in the process of impregnation of carriers with a joint solution of salts of molybdenum, cobalt and nickel stabilized with phosphoric acid, as a rule, the strength of catalyst granules improved in comparison with the carrier. However, it is obviously insufficient for obtaining catalyst granules on carriers №17 - №20 with the strength necessary for loading hydrotreating catalysts into a modern reactor. Some decrease in the share of wide pores may indicate the formation of aggregates of molybdate structures on the walls of the widest pores, especially on the high-percentage catalyst ANM-2/3. In general, the bimodal distribution of pores by radii, characteristic for carriers, was preserved. If citric acid was present in the impregnation solution, the number of wide pores increased, while the pellet strength slightly decreased. This phenomenon was more characteristic of the AKNM-2/9 catalyst containing an increased concentration of active components. Spectral characteristics and phase composition of catalysts AKNM-3/5, ANM-2/3, AKNM/#9 practically did not differ from laboratory analogues (Fig. 4.2, 4.3, 4.7). Approbation of the proposed technological scheme at the UzKFITI SEP during the production of pilot batches of catalysts on a number of carriers allowed to optimize a number of technological parameters that ensure the stability of joint impregnation solutions, time and temperature of impregnation, as well as the uniformity of application of active components [217; 143, 147-148, 244; P.39].

**Table 4.4.**

**Effect of solid component ratio on the texture of catalysts**

| **media no.** | **Impregnation solution stabilizer** | **Catalyst code** | **Weight ratio GOA-M:K:AOA-K** | **$K_{strength}$; kg/mm** | **$D_{cp}$ of pores; nm** | **Fraction of pores with$D_{..}$=7-10 nm;** | **$\sum V_{por}$; $cm^3/g$** |
|---|---|---|---|---|---|---|---|
| №3 | $H_3RO_4$ | AKNM-4/16 | 1.0:0.1:0 | 2.5 | 4.0 | 3.0 | 0.56 |
| №9 | $H_3RO_4$ | AKNM-3/5<br>ANM-2/3 | 1,0:0,1:0,1<br>1,0:0,1:0,1 | 3.4<br>3.5 | 6.8<br>6.3 | 31.2<br>30.4 | 0.53<br>0.51 |
| | $H_3RO_4$ and | ACNM/#9<br>AKNM-2/9 | 1,0:0,1:0,1 | 3.0<br>2.9 | 9.0<br>8.9 | 54.1 | 0.65 |

| media no. | Impregnation solution stabilizer | Catalyst code | Weight ratio GOA-M:K:AOA-K | $K_{strength}$; kg/mm | $D_{cp}$ of pores; nm | Fraction of pores withD =7-10 nm; | $\sum$Vpor; cm$^3$/g |
|---|---|---|---|---|---|---|---|
| | $C_7H_8O_7$ | | 1,0:0,1:0,1 | | | 52.2 | 0.62 |
| №15 | $H_3RO_4$ | ACNM/#15 | 1:0.2:0.3 | 2.1 | 6.7 | 30.3 | 0.50 |
| №16 | $H_3RO_4$ | ACNM/#16 | 1,0:0,5:0,1 | 3.58 | Not op. | Not op. | 0.29 |
| №17 | $H_3RO_4$ | ACNM/#17 | 1,0:0,1:0,5 | 0.62 | Not op. | Not op. | 0.58 |
| №18 | $H_3RO_4$ | ACNM/#18 | 1,0:0,2:0,5 | 1.01 | 6.3 | 27.5 | 0.56 |
| №19 | $H_3RO_4$ | ACNM/#19 | 1,0:0,5:0,5 | 1.22 | 4.9 | 20.5 | 0.47 |
| №20 | $H_3RO_4$ | ACNM/#20 | 9:0:1 | 0.8 | 4.3 | 19.7 | 0.33 |

Thus, the results of the study of a series of catalysts by spectral methods unambiguously proved the influence of the impregnation method on the formation of surface bonds of the Al - O - Mo - O - Ni, Al - O - Mo - O - O - Co and Al - O - Ni - O - O - Mo types. The introduction of cobalt ions into the coordination sphere of molybdate ions in $NiMoO_4$ $xH_2O$ during the synthesis of ACNM catalysts has been proved X-ray diffraction. It is revealed that citric acid in the impregnation solution performs the function of pH regulation, allows increasing the amount of active metals introduced by a single impregnation of the carrier, as well as the amount of structures potentially active in C-S and C=C bond breaking reactions of sulfur-containing hydrocarbons.

## Conclusion

1. On the basis of the regularities of formation of porous structure of the catalyst carrier obtained in this work, the principle of thermochemical activation of large-crystalline aluminum hydroxide - a component of the hydrotreating catalyst carrier - was proposed.
2. A method of regulating the texture of catalyst carriers was developed, providing for peptization of fine crystalline aluminum hydroxide with aqueous solutions of nitric and boric acids, followed by the introduction of enriched Angren kaolin and coarse-dispersed fraction of activated aluminum oxide.
3. The optimal quantitative range of the ratio of fine fractions of fine crystalline aluminum hydroxide and kaolin to the coarse-disperse fraction of activated aluminum oxide necessary for obtaining a carrier with bimodal distribution of mesopores 7-10 nm and transport pores, optimal for oil hydrotreating catalyst, has been revealed.
4. The formation of mixed non-stoichiometric Co-Ni-Mo compounds at the hydrochemical stage of trimetallic catalyst synthesis by a single impregnation method has been demonstrated.
5. The role of mixed non-stoichiometric Co-Ni-Mo compounds as oxide precursors of sulfide phases showing increased hydrogenating and desulfurizing activity towards "difficult sulfur" in mild conditions of hydrotreating of oil fractions has been revealed.
6. The technology of obtaining a new trimetallic catalyst for hydrotreating of oils, combining the positive properties of both aluminocobaltmolybdenum and

aluminonickelmolybdenum catalysts, the use of which will ensure the production of higher quality base oil under relatively mild conditions typical for the Fergana refinery, has been developed.

## List of sources used

**1.** Smirnov V.K., Irisova K.N., Talisman E.L. New catalysts of hydrodehydration of oil fractions and experience of their operation // Catalysis in Industry, Moscow 2003,- № 2.- P. 26-31.

**2.** Saidakhmedov Sh. M. Scientific and technological bases of production of lubricating oils from local hydrocarbon raw materials: Dissertation of Doctor of Technical Sciences. - Tashkent: Academy of Sciences of Uzbekistan, 2005, -70 pp.

**3.** Shteinbrecher A.G., Smolin Yu.B., Obryvalina A.N. High-quality base oils - the basis of a promising product range // Oil refining and petrochemistry, 2005, - № 8. - C.22-23.

**4.** Reznikov V.D., Shipulina E.N. New in foreign classifications of motor oils // Chemistry and technology of fuels and oils, Moscow 2002, - № 4. - C.28-34.

**5.** Lynch, T. R. Process Chemistry of Lubricant Base Stocks / T. R. R. Lynch. - New York: CRC Press, Taylor & Francis Group, 2008. - 369 p. Lynch, T. R. Process Chemistry of Lubricant Base Stocks / T. R. Lynch. R. Lynch. - New York: CRC Press, Taylor & Francis Group 2008, - 369 p.

**6.** Reference. Fuels. Lubricants. Assortment and application under edition of Shkolnikov V.M. Moscow. Techinform. 1999, - C.600.

**7.** Stuffeld R. O. Mobil Catalytic Dewaxing: the Better Alternative to Solvent Dewaxing // Second Russian Refining Roundtable: Vienna, Austria 1998,- November 4- P.124.

**8.** Shkolnikov V.M., Usakova N.A., Stepuro O.S. Catalytic processes of dewaxing in the production of base oils // Chemistry and Technology of Fuels and Oils. Moscow 2000, - №1. - C.23-25.
**9.** Proskurin M.A. Physico-chemical problems of production and application of fuels and lubricants // Oil and Gas, Moscow 1999, - № 2. - C. 59-67
**10.** Yakovlev S.P., Boldinov V.A. Dewaxing and de-oiling with the use of pulsation mixing crystallizer // Chemistry and Technology of Fuels and Oils, 2009,- № 3 P.7-13.
**11.** Kolesnikov A.G., Krylov A.A., Zavalinskaya I.S., Peter Amba Neji Research of regeneration of catalysts for hydrocarbon fractions refining // Oil refining and petrochemistry, 2008, - № 3. - C.26-30.
**12.** Salimov Z.S. Development of ways to solve urgent problems of petrochemistry and oil refining is a priority direction of the Institute // Uzbekistonda neftni kaita ishlashning dolzarb muammolari va moylovchi materiallar ishlab chikarish istikbollari. Respublika ilmiy-tekhnicheskoy konferenciia. Thesislar Tuplami. October 6-7, Tashkent 2005, - P.8-12.
**13.** Akhmetov S.A. Technology of deep processing of oil and gas. - Ufa: Gilem, 2002, - P. 672
**14.** Kitova M.V., Loginova A.N., Vlasov V.G., Tomina N.N., Sharikhina M.A., Lukanov A.A.,. Catalytic dewaxing of weighted diesel fractions. // Chemistry and Technology of Fuels and Oils. - Moscow 2001, No. 1. -C. 16-18.
**15.** Lukića, J. Re-refining of waste mineral insulating oil by extraction with N-methyl-2-pyrrolidone / J. Lukića. Lukića, A. Orlović, M. Spitellerc, J. Jovanović, D. Skalab //

Separation and Purification Technology, 2006, - V. 51. - P. 150-156.
**16.** Smirnov V.K., Gantsev V.A., Kapustin V.M. New catalysts for hydrodehydration of oil fractions // Chemistry and Technology of Fuels and Oils, Moscow 2002, - № 3. - C. 3-7.
**17.** Safronova T.N. Hydrodesulfurization and hydrogenation of components of oil fractions on Ni(Co)Mo(W)/$Al_2O_3$ catalysts. // Dissertation of Candidate of Chemical Sciences. Samara 2014, - P. - 164.
**18.** Aliev R.R., Yolshin A.I., Reznichenko I.D. Problems and criteria for selecting catalysts for hydrotreating of oil fractions // Chemistry and Technology of Fuels and Oils - Moscow 2002, - No. 2. - C. 16-18.
**19.** Smirnov V.K., Gantsev V.A., Sukhorukov A.Ya. et al. Deep hydrodehydrogenation of vacuum gasoil at the installation G-43-107 // Chemistry and Technology of Fuels and Oils - Moscow 2001, - № 4 - P. 4-6. 4-6.
**20.** Smirnov V.K., Irisova K.N. et al. New catalysts for soft hydrocracking of vacuum distillate // Chemistry and Technology of Fuels and Oils, Moscow 1999, -#2. -C. 18-20.
**21.** Elshin A.I., Anufriev V.I. et al. Increase of operation efficiency of hydrotreating unit L-24-6 // Chemistry and technology of fuels and oils, Moscow 2000. - №3, - C. 36-38
**22.** Safronova T.N., Tomina N.N., Pimerzin A.A. Hydrotreating of selective purification raffinate on Ni6-PW12/Al2O3 catalyst. // In the collection: "Theses of reports of scientific and technological symposium "Refining: catalysts and hydroprocesses". Pushkin. St. Petersburg 2014,- P.199-200.

**23.** Saidakhmedov S.M., Tojiev E.T. State and prospects of oil refining development in Uzbekistan // Chemistry and Technology of Fuels and Oils, Moscow 1996,- No.4. - C.3-5.
**24.** Ergashev M.M. Main achievements and prospects of development of Fergana oil refinery // Collection of theses of the Republican scientific and technical conference, Tashkent 2005. -C.71-73
**25.** Khairutdinov I.R., Sayfullin N.R., et al. Propane-butane deasphaltization of tar // Chemistry and technology of fuels and oils, Moscow 1999, - №3.- P.14-15.
**26.** Levina L.A., Zelentsov Y.N., Yolshin A.I. et al. "Catalyst System for Hydrotreating of Base Oils" // Chemistry and Technology of Fuels and Oils, Moscow 2003, № 4, -C. 14-15.
**27.** Antonov S.A. Safronova T.N., Pimerzin A.A. Chemical composition of oil fractions of sulfuric oils and their rational processing // In Proceedings of the All-Russian Scientific and Practical Conference "New Technologies - oil and gas region", Tyumen 2013, - P. 84-85.
**28.** Sochevko T.I., Pakhomov M.D., Falkovich M.I., Evtushenko V.M. Base and commercial oils. Factors of quality improvement // Chemistry and technology of fuels and oils. - Moscow 2002, - №2. - C.37-39.
**29.** Saidullaev B.T., Ergashev M.M., Musaeva G.H. et al. Technology of M-10 DM oil production on the basis of oil distillates of FNPZ // Uzbek Journal of Oil and Gas - Tashkent 2006.- No.3.- P.34-36.
**30.** Ergashev M.M., Saidullaev B.T., Musaeva G.H. et al. Obtaining all seasonal oil for high-force engines from high-

sulfur oil raw materials // Uzbek Journal of Oil and Gas - Tashkent 2006, - No.4. - C.23-24.
**31.** Catalyst for hydrotreating of oil fractions and selective purification raffinates and method of its preparation Tomina N.N., Pimerzin A.A. et al. Patent of the Russian Federation № 2497585 - 10.11.2015.
**32.** Radchenko L.A., Chesnokov A.A. New generation oils for hydraulic systems of industrial equipment // Chemistry and technology of fuels and oils, Moscow 2003,- №3. - C. 33-35.
**33.** Nigmatullin R.G., Batyrov N.A., Vorobyev A.A., Olkov P.L., Aznabaev Sh.T. Selective hydrocracking of partially dewaxed raffinates // Chemistry and Technology of Fuels and Oils, Moscow 2000,-[1]1. - C.26- 27
**34.** Salimov Z.S., Kadyrov I., Saidakhmedov Sh. Polyfunctional catalysts and hydrogenation processes of oil refining. - Tashkent: Fan, 2000. - C. 109.
**35.** Abidova M.F., Iskandarov S.A., Ibragimov K.A. Activity of hydrocracking catalysts and technology of their preparation // Uzbek Journal of Oil and Gas, Tashkent 2005, - No. 1, - P. P. 30-32.
**36.** Kadyrov I., Salimov Z.S., Solomov Y. et al. Hydrotreatment of light straight-run gasoline fractions of oil // DAN RUz . - Tashkent 2002, - №4. - C.61-63.
**37.** Kadyrov I. Perspective catalytic processes of oil refining // Uzbek Journal of Oil and Gas, Tashkent 2003, - No.1, - P.51-54.
**38.** Abidova M.F., Iskandarov Sh.A. et al. Selection of catalysts for deepening of oil feedstock processing // Uzbek Journal of Oil and Gas, Tashkent 2005,- № 3.- P. 44-46. 44-46.

**39.** Abidova M.F., Saidakhmedov S.M., Karimkulova M.P. State of development of hydroprocessing of residual oil feedstock (message 1) // Uzbek Journal of Oil and Gas. - Tashkent. 2003. - №4. - C.26-28.

**40.** Abidova M.F., Saidakhmedov S.M., Karimkulova M.P. State of development of hydroprocessing of residual oil feedstock (message 2) // Uzbek Journal of Oil and Gas, Tashkent 2004,- No.1. - C.28-30.

**41.** Saidakhmedov S.M., Ibragimova G.S., Iskandarov Sh.A., Abidova M.F. Effective catalyst for hydrodehydration of oil distillate // Uzbek Journal of Oil and Gas, Tashkent 2001, - No. 1. - C.33-34.

**42.** Khalafova I.A., Huseynova A.D., Poladov F.M., Yunusov S.G. Study of the process of catalytic refining of gasoline fraction of coking // Chemistry and Technology of Fuels and Oils, 2012. №4.-C.24-26.

**43.** Sayfidinov B. M., Nigmatullin V. R., Sharipov A. H., Nigmatullin. I. R. Purification of light fractions of South Uzbek oils from sulfur compounds // Chemistry and Technology of Fuels and Oils, 2012, - № 5 -C. 15-17.

**44.** Nefedov B.K., Antonov A.E., Sabitova V.M., A new way to intensify the process of hydrodesulfurization of straight-run diesel fractions // Catalysis in Industry, Moscow 2004, - № 5. -C. 12-16

**45.** Tkachev S.M. Technology of oil and gas processing. Processes of deep processing of oil and oil fractions // Training and methodical complex. Novopolotsk 2006, - P. 344.

**46.** Nigmatullin V.R., Mukhametova R.R., Nigmatullin I.R. Oxidative desulfurization of oil distillates // Chemistry and technology of fuels and oils, 2008, -№1. -C. 10-11.

**47.** Pavlov, I.V.; Ponyaev, L.A.; Kazulina, E.V.; Gusakova, J.Yu. Application of hydrogenation processes in the production of base oils (in Russian) // Chemistry and technology of fuels and oils, 2008, - № 2, - P.27-28.

**48.** Sharipov A.H., Nigmatullin I.R., Nigmatullin V.R. Purification of oil fractions from sulfides // Chemistry and Technology of Fuels and Oils, 2009, - № 2. -C. 14-19.

**49.** Kayukova G.P., Petrov S.M., Romanov G.V. Application of hydrogenation processes for obtaining white oils from heavy oil of the Ashalchinskoye field // Chemistry and Technology of Fuels and Oils 2012,- № 4. -C. 9-15.

**50.** Yunusov M.P., Molodozhenyuk T.B., Ergashev M.M. et al. Investigation of the protective layer system for hydrotreating process of oil distillates from Uzbek oils // Chemical Industry, St. Petersburg 2007, - No. 2. - C. 55-61.

**51.** Method of obtaining a protective layer catalyst. Yunusov M.P., Jalalova Sh.B., Isayeva N.F., Molodozhenyuk T.B., Mirzaeva E.I., Mahkamov H.M. // Intellectual Property Agency of the Republic of Uzbekistan, No. IAP 05063, 2015.

**52.** Catalyst for hydrotreating of the upper layer and method of its preparation Polunkin Y.M., Shragina G.M. et al. // Patent of the Russian Federation № 2235588. - 10.09.2004.

**53.** Protective layer catalyst for hydrotreating of oil fractions. Reznichenko I. D., Aliev R. R. and others; - Russian Federation patent No. 2319543. - 20.03.2008.

**54.** Tarakanov G.V., Nurakhmedova A.F. Multilayer catalyst systems for hydrodehydration of oil fractions //

Chemistry and Technology of Fuels and Oils. № 6. 2007.- C.48-51.

**55.** Kamalova D.N. Research of the basic parameters of equipment in the production of road surfaces at the Tashkent asphalt-bitumen plant. Diss. Master's degree. Tashkent 2014. - C.77

**56.** Nabiev A.B., Abdurakhimov S.A. Classification of local oils from the position of their fluidity // Chemistry and Chemical Technology - Tashkent. - 2009. - №4. - C.63-65.

**57.** Safarov B.J., Khayitov R.R., Muradov M.N. Composition and properties of resins contained in the oil of Uzbek fields // Chemistry and Chemical Technology - Tashkent. - 2009. - №2. - C.59-61

**58.** Konovalchikov O.D., Khavkin V.A., Gulyaeva L.A., Krasilnikova L.A., Misko O.M., Bychkova D.M., Loschenkova I.N., Druzhinin O.A.. Catalysts of destructive hydrotreating of heavy diesel distillates // Catalysis in industry. - Moscow, 2004, № 6, P. 21-26.

**59.** Kurganov V.M., Papusha L.V. et al. Hydrocracking Process in the Scheme of Oil Production // Chemistry and Technology of Fuels and Oils. - Moscow. 1999. - №3, - C. 13.

**60.** Smirnov V.K., Irisova K.N., Talisman E.L. et al. Hydrotreatment of oil raffinates on catalyst RK-438 // Oil refining and petrochemistry. №3. 2008.-C.22-25

**61.** Berg G.A., Khabibullin S.G. Catalytic hydrodehydration of oil residues. L. Khimiya.1986. -C 192.

**62.** Krasilnikova G.M., Spiridonov S.E., Baibursky V.L. et al. Highly effective catalyst for hydrodehydrogenation // Chemistry and Technology of Fuels and Oils.- Moscow. 1999. №3. -C. 21.

**63.** Vezirov R.R. Visbreaking - time-tested technologies. // Chemistry and technology of fuels and oils. 2010. №6. -C..3-8.

**64.** Velichkina, L.M.; Gossen, L.P. Areas of application of new catalysts of resource-saving and environmentally friendly technologies in oil refining and petrochemistry (in Russian) // Neftekhimiya. 2008.№ 3.-C. 31-37.

**65.** Starkova N.N., Shuverov V.M., Ryabov V.G., Yunusov Sh.M. Characterization of raw materials for obtaining high-index base oils // Chemistry and technology of fuels and oils. - Moscow. 2001 -№ 3.- C. 36-37.

**66.** Nefedov B.K., Radchenko E.D., Aliev R.R. Catalysts of processes of advanced oil refining. M., Khimiya, 1992, -C. 272

**67.** Kuliev R.Sh. Two-column scheme of selective purification of oil raw materials // Chemistry and technology of fuels and oils.- Moscow. 2005 .-№1.-C.22-23

**68.** Kotov S.V., Oltyrev A.G., Shabalina T.N. et al. Improving the quality of oil distillates // Chemistry and technology of fuels and oils.- Moscow. 1999. -C. 10-12

**69.** Parunin P.D., Lysikov A.I., Okunev A.G., Parkhomchuk E.V., Polukhin A.V. Semeikina V.S. Study of 3 D - structured catalysts in the process of hydroprocessing of heavy oil // Scientific and Technological Symposium Petroleum Refining: Catalysts and Hydroprocesses . May 20-23, 2014, Pushkin, St. Petersburg, Novosibirsk 2014.- P. 155-156.

**70.** Reznichenko I.D. Development of hydrotreating catalysts with improved ecological properties on the basis of modified carriers: Cand. Sci. (Techn.) Dissertation. - Ufa:UGNTU, 2007. - C.117.

**71.** Levin O.V. Improvement of catalysts for hydrotreating of gasoline and diesel fractions by optimization of the carrier texture: Cand.Chem.Sci. (in Russian). - Kazan: 2002.- P.135.
**72.** Lamberov AA, Romanova R.G., Levin O.V., Egorova S.R. Influence of structural characteristics of the carrier on the activity of hydrotreating catalysts // Proc. of Russian Conf. "Actual problems of petrochemistry" April 17-20, 2001 - Moscow. 2001. - C.153.
**73.** Lamberov A.A., Levin O.V., Egorova S. R., Gelmanov H.H.. Influence of precipitation temperature on the texture of cool precipitation aluminum hydroxide // XV Intern. Conf. On Chemical Reactors. Helsinki. Finland. June 5-8. 2001. - H. 305-307
**74.** Kulko EV, Ivanova AS, Litvak GS, Kryukova GN, Tsybulia SV Obtaining phase-homogeneous aluminum oxides and study of their microstructure and texture // Kinetics and Catalysis 2004. vol. 45. № 5, -C.-754-762.
**75.** Catalyst for hydrotreating of oil fractions and method of its preparation. Reznichenko I.D., Tselutina M.I., Yolshin A.I., Aliev R.O., Volchatov L.G., Bocharov A.P., Kuks I.V., Trofimova M.. V., Andreeva T.I. // Patent RF, № 2306978.- 27.09.2007 Bulletin № 27
**76.** Catalyst, method of its preparation (variants) and process of hydrodesulfurization of diesel fractions. Ismagilov Z., Shikina NV, Yashnik SA, Rogov VA Kerzhentsev MA, Parmon VN, // Patent RF, 2342994.- 10.01.2009.
**77.** Catalyst for hydrodesulfurization of diesel fraction and method of its preparation. Klimov O. V., Kodenev E. G. G., Echevsky G. V., Bukhtiyarova G. A., Polunkin Ya. V. // Patent of the Russian Federation, 2313392. - 27.12.2007г.

**78.** Method of obtaining catalyst for hydrotreating of oil fractions. Nasirov R. K. // Patent RF, №2074025 - 27.02.1997.
**79.** Method of preparation of catalyst for hydrotreating of oil distillates. Dianova S.A., Kovalchuk N.A., Mund N.S., Karelsky V.V. // Patent RF, № 2179886 - 27.02.2002.
**80.** Sidelkovskaya V.G. et al. Investigation of aluminonickel-molybdenum catalysts based on natural clays // KK 1990. Vol. 31, issue 6 - P.-1414-1419.
**81.** Rabinovich G.L. Shavandin Yu.A., Polotskaya G.E., Zharkov B.B. Method of preparation of catalyst for hydrotreating of oil fractions // Patent RF, № 2089290 - 1997.
**82.** Artukova G.Sh., Ibragimov E.A. et al. Synthesis of hydrotreating catalyst using local raw materials // Journal of Applied Chemistry. 2001.vol.74. vol.6.-P.939-942.
**83.** Jalalova S.B., Saidakhmedov S.M.,. Molodozhenyuk T.B., Yunusov M.P. Structure of hydrocarbon feedstock hydrotreating catalysts obtained on alumokaolin carrier // Journal "Oil and Gas of Uzbekistan" № 2. 2002. C. 28-30.
**84.** Chukin G.D. Structure of aluminum oxide and hydrodesulfurization catalysts. Mechanisms of reactions // Moscow 2010. - C.287
**85.** Leonova K.A., Klimov O.V., Pereyma V.Yu., Dick P.P., Budukva S.V., Uvarkina D.D., Noskov A.S. Effect of silicon content in aluminosilicate carriers on the activity of Co-Mo hydrotreating catalysts in the hydrogenolysis of thiophene // Scientific and Technological Symposium Petroleum Refining: Catalysts and Hydroprocesses. May 20-23, 2014, Pushkin, St. Petersburg, Novosibirsk 2014. - C. 57-58.

**86.** Maity S.K. Alumina-silica binary mixed oxide used as support of catalysts for hydrotreating of Maya heavy crude // Appl . Catal A 250:231-238 Copyright (C) 2013 American Chemical 29.
**87.** Dibenzothiophene and 4, 6-dimethyldibenzo-thiophene: Effect of an acid component on the activity of a sul® ded NiMo on alumina catalyst. Applied Catalysis A: General, 169. - C.343-353.
**88.** Usman U.,Takaki M., Kubota T., Okamoto Y. "Effect of boron addition on a $MoO_3/Al_2O_3$ catalyst" // Applied Catalysis A: General, 286, 2005.-C.148-154.
**89.** Kubota T.; Araki Y.; Ishida K.; Okamoto Y. "The effect of boron addition on the hydrodesulfurization activity of $MoS_2/Al_{(2)}O_3$ and Co-$MoS_{(2)}/Al_{(2)}O_3$ catalysts" // Journal of Catalysis, 227, 2004.-P. 523-529.
**90.** Kubota T., Hiromitsu I., Okamoto Y. "Effect of boron addition on the surface structure of Co-$Mo/Al_2O_3$ catalysts" // Journal of Catalysis; 2007; 247;-P.78-85.
**91.** Ferdous D., Dalai A. K., & Adjaye J. Hydrodenitrogenation and Hydrodesulfurization of Heavy Gas Oil Using NiMo/Al2O3 Catalyst Containing Boron: Experimental and Kinetic Studies. Industrial & Engineering Chemistry Research, 45, 2006. - P. 544-552.
**92.** Saih Y., & Segawa K. Catalytic activity of CoMo catalysts supported on boron-modified alumina for the hydrodesulphurization of dibenzothiophene and 4,6 dimethyldibenzothiophene. Applied Catalysis A: General, 353, 2009.-P. 258-265.
**93.** Kim H., Lee J.J., Moon S.H.. "Hydrodesulfurization of dibenzothiophene compounds using fluorinated $NiMo/Al_2O_3$

catalysts" // Applied Catalysis B: Environmental; 2003; 44; 287-299.

**94.** Ding L., Zhang Z., Zheng Y., Ring Z., Chen J. "Effect of fluorine and boron modification on the HDS, HDN and HDA activity of hydrotreating catalysts" // Applied Catalysis A: General, 2006, 301. -P.241-250.

**95.** Ferdous D., Dalai A. K., & Adjaye J. (2004). A series of NiMo/Al2O3 catalysts containing boron and phosphorus. Applied Catalysis A: General, 260.-P.137-151.

**96.** Ferdous D., Dalai A. K., & Adjaye J. (2005). X-ray absorption near edge structure and X-ray photo electron spectroscopy analyses of NiMo/Al2O3 catalysts containing boron and phosphorus. Journal of Molecular Catalysis A: Chemical, 234.-P. 169-179.

**97.** Alekseenko L.N., Landau M.V., Nefedov B.K. On the ratio of hydrogenating and hydrodesulfurizing activity of aluminonickelmolybdenum catalysts // Kinetics and Catalysis 1984. vol.25. vol.2. -C.492-496.

**98.** A.V.Vysotsky, I.F.Sarapulova, V.A.Yaskina. About acidity of aluminonickel-molybdenum hydrotreating catalysts promoted by zeolites // Kinetics and Catalysis. T. 33. №1, 1992, 197-204.

**99.** Jalalova S.B., Saidakhmedov S.M., Molodozhenyuk T.B., Ibragimov K.A., Yunusov M.P. // Study of the genesis of surface centers in the synthesis of catalysts for hydrotreatment of hydrocarbon feedstock // J. "Chemical Industry", St. Petersburg 2003,- № 2. -C.3-8.

**100.** Lavrenov A.V., Basova I.A., Kazakov M.O. et al. Catalysts on the basis of anion-modified metal oxides for obtaining environmentally friendly components of motor fuels // Russian Chemical Journal. 2007. Vol. II. No. 4. p. p. 75-84.

**101.** Korobochkin V.V., Gorlushko D.A. Technology of catalysts. Part 1. Methods of preparation of catalysts. Tomsk 2013. - C. 79

**102.** Moreno B., Chinarro E., Colomer M. T., & Jurado J. R. Combustion Synthesis and Electrical Behavior of Nanometric B-$NiMoO_4$. Journal of Physical Chemistry C, 114, 2010.-C.4251-4257.

**103.** Juanjuan L., Zhiqiang X., Weikun L., Jinbao Z., Binghui Ch., Xiaodong Y., Weiping F. Hydrodemetallation (HDM) of nickel-5, 10,15,20-tetraphenylporphyrin (Ni-TPP) over NiMo/c-$Al_2O_3$ cyst prepared by one-pot method with controlled precipitation of the components // Fuel 97. 2012 - P.504-511

**104.** Dryaglin Yu.Yu., Solmanov P.S. Studies of methods of introducing cobalt into CoO-$MoO_3$/γ-$Al_2O_3$ hydrotreating catalysts // In Proceedings of the XXI Mendeleev Conference of Young Scientists. - Dubna, 2011. - C. 61.

**105.** Rana M.S. Effect of catalyst preparation and support composition on hydrodesulfurization of dibenzothiophene and Maya crude oil // Fuel 86.-2007.-P.1254-1262

**106.** Rayo P., Rana M., Ramirez J., Ancheyta J., & Aguilarelguezabal. Effect of the preparation method on the structural stability and hydrodesulfurization activity of NiMo/SBA-15 catalysts // Catalysis Today, 130.-2008.-P.283-291

**107.** Landau MV, Mikhailov VI, Tsisun EL, Samgina TY, Chukin GD, Vinogradova OV, Nefedov BK, Slinkin AA Synthesis and catalytic properties of sulfidation products of ternary oxide compounds Ni Mo Al in hydrogenation and hydrodesulfination reactions // KK vol. 27. Vol.4. 1986. - .C.931-940

**108.** Landau M.V., Alekseenko L.N., Vinogradova O.V. et al. Effect of deposition conditions and heat treatment of nickel molybdate on the composition and structure of the resulting product // Kinetics and Catalysis. №4. 1989.- C. 933-938.

**109.** Victor Costa. Comprehension du role des additifs du type glycol surl'amelioration des performances des catalyseurs d'hydrotraitement // du diplome de doctorat. Catalysis. Universite Claude Bernard - Lyon I, 2008. French. - P.263.

**110.** Sidelkovskaya V.G. Features of formation of active phases in aluminonickelmolybdenum catalysts based on various aluminum hydroxides // Kinetics and Catalysis. Vol. 31, issue 5. 1990. - C.1264-1267.

**111.** Agievsky D.A., Kvashonkin V.I., Pavlova L.I., Chukin G.D.. Mikhailov V.I., Surin S.A., Landau M.V., Nefedov B.K. Study of oxide precursors of active structures in Al-Ni-Mo hydrogenation catalysts by extractive separation of components I // Kinetics and Catalysis, 1986. vol. 27, №1 - C.178-185.

**112.** Chukin G.D., Mikhailov V.I., Surin S.A. et al. Study of oxide precursors of active structures in Al-Ni-Mo hydrogenation catalysts by extractive separation of components II // Kinetics and Catalysis - Moscow. 1986. T. 27. -№ 1. - C. 186-193.

**113.** Goncharova O.I., Davydov A.A., Yurieva T.M. IR spectroscopic manifestation of polymolybdenum compounds on the surface of molybdenum-aluminum catalysts // Kinetics and Catalysis. 1984, Vol. 25, No. 1.-P. 152-158.

**114.** Spozhakina AA, Kostova, NG, Tsolovski IA, Shopov DM Synthesis and properties of hydrotreating catalysts. 2.The state of molybdenum in aluminomolybdenum catalysts and their catalytic properties in the hydrogenolysis of thiophene // Kinetics and Catalysis. 1982. т. 23. vol.2.-P.456-461.
**115.** G.D. Chukin. Study of structure formation and active phases in hydrotreating ANMC with additives of amorphous and crystalline silicates // Kinetics and Catalysis.V.31. vol. 1990.- P.503.
**116.** Surin S.A. Mikhailov V.I., Sidelkovskaya V.G., Chukin G.D., Nefedov B.K. et al. Spectroscopic study of aluminonickel-molybdenum catalysts based on various modifications of $Al_2O_3$ // Kinetics and Catalysis. T. 28. №4. 1987. -C. 943-947
**117.** Surin S.A., Chukin G.D., Sidelkovskaya V.G. et al. Influence of the order of application of active components on the structure, phase composition and activity of ANM catalysts // Kinetics and Catalysis Th. 29. V.2. 1988 c. 443-446.
**118.** Gazimzyanov N.R., Mikhailov V.I., Zadko I.I.. Study of distribution of active components in aluminonickel-molybdenum catalysts obtained by single impregnation // KK 1992. vol.33, issue 4.-P. 915-921.
**119.** Iwamoto R., Grimblot J. Influence of Phosphorus on the Properties of Alumina-Based Hydrotreating Catalysts // Advances in Catalysis 44. 1999. - P.417-503.
**120.** Atanasova P., Tabakova T., Vladov C., Halachev T., & Agudo A. L. Effect of phosphorus concentration and method of preparation on the structure of the oxide form of

phosphorus-nickel-tungsten/alumina hydrotreating catalysts // Applied Catalysis A: General 161. 1997.-P. 105-119.

**121.** Livage C., Hynaux A., Marrot J., Nogues M., Férey G. Solution process for the synthesis of the "high-pressure" phase $CoMoO_4$ and X-ray single crystal resolution // J. Mater. Mater. Chem. 2002. 12. -P. 1423-1425.

**122.** Maity S., Flores G., Ancheyta J., Rana M. "Effect of preparation methods and content of phosphorus on hydrotreating activity" // Catalysis Today. 2008; 130.-P. 374-381

**123.** Maity S. K., Ancheyta J., Rana M. S., & Rayo P. Effect of phosphorus on activity of hydrotreating catalyst of Maya heavy crude // Catalysis Today. 2005.-P. 10942-10948.

**124.** Patricia R, Jorge R, Pablo T, Gustavo M, Samir K. M., Jorge A. Hydrodesulfurization and hydrocracking of Maya crude with P-modified NiMo/$Al_2O_3$ catalysts // Fuel 100. 2012.-P. 34-42.

**125.** Xiang C., Chai Y., Fan J., & Liu C. Effect of phosphorus on the hydrodesulfurization and hydrodenitrogenation performance of presulfided NiMo / $Al_2O_3$ catalyst // Journal of Fuel Chemistry and Technology. 39. 2011.-P. 355-360.

**126.** Pouler O., Hubaut R., Kasztelan S., Grimblot J. Experimental evidences of the direct influence of phosphorus on activity and selectivity of a $MoS_2/\gamma\text{-}Al_2O_3$ hydrotreating catalyst // 4th th Workshop Hydrotreat, Louvainla-Neuve, Nov. 21-22, 1991 // Bull. Soc. Chim. Belg. 1991. 100. N11-12 . -P.. 857-863.

**127.** Plazenet, G., E. Payen, and J. Lynch, "Cobalt-Molybdenum Interaction in Oxidic Precursors of Zeolite

Supported HDS Catalysts," // Phys. Chem. Chem. Phys. 4. 2002. - P. 3924-3926

**128.** Sundaramurthy V., Dalai A. K., Adjaye J. Effect of phosphorus addition on the hydrotreating activity of NiMo/ $Al_2O_3$ carbide catalyst // Catalysis Today. 125. 2007. - P.239-247.

**129.** Sundaramurthy, V., Dalai, A. K., Adjaye, J. The effect of phosphorus on hydrotreating property of NiMo/γ- $Al_2O_3$ nitride catalyst // Applied Catalysis A: General. 335. 2008.- P. 204-210.

**130.** Maksimov N.M., Dryaglin Y.Yu., Solmanov P.S., Tomina N.N. Effect of modification by boron and phosphorus compounds of $Co_6PMo_{12}/Al_2O_3$ catalysts on their activity // In Proceedings of the All-Russian Scientific Conference "Hydrocarbon Feedstock Processing. Complex Solutions" - Samara, 2012 - P.81.

**131.** Eremina Yu.V. Study of features of reactions of hydrodesulfination and hydrogenation of components of diesel fractions on molybdenum-containing catalysts. diss. candidate of chemical sciences. - Samara:ASU, 2006. - C.136.

**132.** Sun M., Nicosia D., Prins R. The effects of fluorine, phosphate and chelating agents on hydrotreating catalysts and catalysis // Catal. Today. - 2003. - V.86. - P. 173-189.

**133.** Klimov O.V., Fedotov M.A., Pashigreva A.V., Budukva S.V., Kirichenko E.N., Bukhtiyarova G.A., Noskov A.S. Complex compounds formed in solution from ammonium paramolybdate, orthophosphoric acid, cobalt or nickel nitrates and urea, and prepared on their basis catalysts hydrotreatment of diesel fuel // Kinetics and Catalysis. 2009. T. 50. № 6. -C. 903-909.

**134.** Shafi R., Siddiqui M.R., Hutchings, G.J., Derouane E.G., Kozhevnikov, I.V.. "Heteropoly acid precursor to a catalyst for dibenzothiophene hydrodesulfurization" // Applied Catalysis A: General; 2000. 204.-P. 251-256

**135.** Ishutenko D.I., Minaev P.P., Nikulshin P.A., Konovalov V.V. HDS and HYD activities of sulfide catalysts synthesized with heteropolycompounds Anderson's structure // 5th International symposium on the molecular aspects of catalysis by sulfides, Denmark. 2010.- P. 254.

**136.** Nikul'shin P.A., Mozhaev A.V., Ishutenko D.I., Minaev P.P., Lyashenko A.I., Pimerzin A.A. "Influence of the composition and morphology of nanosized transition metal sulfides prepared using the Anderson-type heteropoly compounds $[X(OH)_{(6)}Mo_{(6)})Mo_{(6)})O_{18}]_{(n)}$)- (X = Co, Ni, Mn, Zn) and $[Co_2Mo_{10}O_{38}H_4]_{(6)}$)- on their catalytic properties" // Kinetics and Catalysis, 2012,53.-P. 620-631.

**137.** Nikulshin P.A., Tomina N.N., Pimerzin A.A., Stakheev A.Y., Mashkovsky I.S., Kogan V.M. "Effect of the second metal of Anderson type heteropolycompounds on hydrogenation and hydrodesulphurization properties of $XMo_6(S)/Al_2O_3$ and $Ni_{(3)}$-$XMo_6(S)/Al_2O_{(3)}$ catalysts" // Applied Catalysis A: General; 2011.No. 393.-P. 146-152.

**138.** Pettiti I., Botto I., Cabello C., Colonna S., Faticanti M., Minelli G., Porta P., Thomas H. "Anderson-type heteropolyoxomolybdates in catalysis: 2. EXAFS study on γ-$Al_2O_3$-supported Mo, Co and Ni sulfided phases as HDS catalysts" // Applied Catalysis A: General. 2001. 220.-P. 113-121.

**139.** Ishutenko D.I. Activity of hydrodesulfurization catalysts based on some heteropoly compounds of molybdenum 6-th row in the reaction of hydrogenolysis of thiophene // In Collection: Theses of reports of the 63rd student scientific conference "Oil and Gas - 2009", Moscow 2009. -C. 25.

**140.** Ishutenko D.I. Activity of hydrodesulfination catalysts based on some heteropoly compounds of molybdenum 12 in the reaction of hydrogenolysis of thiophene // In Proc. of XVIII Mendeleev Conference of Young Scientists. - Belgorod, 2008. -C.93-94.

**141.** Denisov A.V., Dryaglin Y.Yu., Solmanov P.S. Investigation of hydrotreating catalysts prepared on the basis of GPC and GPC molybdenum 12 row Keggin structure in the model reaction of hydrogenolysis of thiophene // In Proceedings of the XX Mendeleev Conference of Young Scientists. - Arkhangelsk, 2010. - C. 56.

**142.** Denisov A.V., Dryaglin Y.Yu., Solmanov P.S. Synthesis and study of highly active hydrotreating catalysts based on heteropolyacids of molybdenum 12-row Keggin structure // In Proceedings of the XX Mendeleev Conference of Young Scientists - Arkhangelsk, 2010. - C. 58.

**143.** Maksimov N.M., Solmanov P.S., Dryaglin Y.Y., Tomina N.N., Pimerzin A.A. $XMo_{12}$ - heteropoly compounds - precursors of hydrodesulfurization catalysts. GDS, GDA and GID activity // In the collection "Abstracts of the 6th International Symposium "Molecular Aspects of Catalysis by Sulfides". - France, 2013. - C. 92.

**144.** Tomina N.N., Maximov N.M., Dryaglin Yu Yu, Solmanov P.S., Denisov A.V. Comparative activity of sulfide catalysts based on heteropoly compounds 12 series // In Proceedings of the "Eurasian Symposium on Innovations in Catalysis and Electrochemistry". - Alma-Ata, 2010. - C. 163.

**145.** Lamonier C.; Martin C.; Mazurelle J.; Harlé V.; Guillaume D.; Payen E. "Molybdocobaltate cobalt salts: New starting materials for hydrotreating catalysts" // Applied Catalysis B: Environmental; 2007. 70.-P. 548-556.

**146.** Mazurelle J.; Lamonier C.; Lancelot C.; Payen E.; Pichon C.; Guillaume D. "Use of the cobalt salt of the heteropolyanion $[Co_2Mo_{10}O_{38}H_4]_{(6)}$- for the preparation of CoMo HDS catalysts supported on $Al_{(2)}O_3$, $TiO_2$ and $ZrO_2$" // Catalysis Today; 2008; 130.-P. 41-49.

**147.** Mozhaev A.A.; Nikul'shin P.A.; Pimerzin A.A.; Konovalov V.V.; Pimerzin A.A.. "Activity of Co(Ni)MoS/$Al_2O_3$ catalysts, derived from cobalt(nickel) salts of $H_6[Co_2Mo_{10}O_{38}H_4]$, in hydrogenolysis of thiophene and hydrogen treatment of diesel fraction" // Petroleum chemistry; 2012.№ 52.-P. 45-53.

**148.** Moshaev A.V., Nikulshin P.A., Konovalov V.V., Eremina Yu. V. and Pimerzin A.A. Use of $Co_2Mo_{10}$-heteropolycompounds for preparation of highly active CoMoS phase of type II // 5th International symposium on the molecular aspects of catalysis by sulfides, Denmark, 2010.- P. P. 258.

**149.** Mozhaev A.V., Nikulshin P.A., Konovalov V.V., Pimerzin A.A. Investigation of the influence of Co(Ni)/Mo ratio in catalysts Co(Ni)(Chel)$Co_2Mo_{10}$GPC/$Al_2O_3$ on the morphology of the active phase and catalytic behavior //

Abstracts of the Conference of Young Scientists in Petrochemistry. Zvenigorod. 2011 -C. 69.

**150.** Nikulshin P.A., Moshaev A.V., Konovalov V.V., Tomina N.N. and Pimerzin A.A. Simultaneous use of $Co_2Mo_{10}$-heteropolycompounds and chelating agents for synthesis of highly active hydrotreating catalysts // EuropaCatIX, Spain. 2009. CD. -P.4-28.

**151.** Pashigreva A.V. "Co-Mo catalysts of deep hydrotreating of diesel fractions prepared through the stage of synthesis of bimetallic compounds": Author's abstract of dissertation of Candidate of Chemical Sciences. Novosibirsk. 2009. - 16

**152.** Klimov O.V., Pashigreva A.V., Leonova K.A., Bukhtiyarova G.A., Budukva S.V., Noskov A.S. Bimetallic Co-Mo-complexes with optimal localization on the support surface: a way for highly active hydrodesulfurization catalysts preparation for different petroleum distillates // Stud. Surf. Sci. Catal. 175. 2010.-P 509-512.

**153.** Pashigreva A.V., Bukhiyarova G.A., Klimov O.V., Noskov A.S. Activity and sulfidation behavior of $CoMo/Al_2O_3$ hydrotreating catalyst: effect of drying condition // 14th International Congress on Catalysis: Abstracts, July 13-18.Seoul, S. Korea, 2008.- P. P. 279.

**154.** Pashigreva A.V., Klimov O.V., Bukhtiyarova G.A. at al. The superior activity of the CoMo hydrotreating catalysts prepared using citric acid: what's the reason? // $10^{th}$ International Symposium "Scientific Bases for the Preparation of Heterogeneous Catalysts" - 2010.-C. 109-116.

**155.** Pashigreva A. V., Bukhtiyarova G. a., Klimov O. V., Chesalov Y. V., Chesalov Y. A., Litvak G. S., & Noskov a.

S. Activity and sulfidation behavior of the CoMo/Al2O3 hydrotreating catalyst // The effect of drying conditions. Catalysis Today. 149. -2010.-P 19-27.

**156.** Pena L., Valencia D., Klimova T. CoMo / SBA-15 catalysts prepared with EDTA and citric acid and their performance in hydrodesulfurization of dibenzothiophene // AppliedCatalysis B: Environmental. 2014. V.147.-P.879-887.

**157.** Pashigreva A.V., Bukhtiyarova G.A., Klimov O.V. Influence of synthesis, drying and sulfidation conditions on properties of hydrotreating catalysts prepared using bimetallic Co-Mo complex compounds // Chemistry under the sign "Sigma": Mat. All-Russian Scientific Young Conference, May 18-23, 2008, Omsk, -2008, - P. P. 175-176.

**158.** Hai-liang Y., Tong-na Z., Yun-qi L., Yong-ming C., Chen-guang L.. Study on the structure of active phase in NiMoP impregnation solution using Laser Raman spectroscopy. II. Effect of organic additives // Journal of Fuel Chemistry and Technology. -2011. -V.39(2). -P.109-114.

**159.** Vdovina T.N., Bely A.S., Shkuropat S.A., Startsev A.N., Dupliakin V.K.. Distribution of the active component in pores of different sizes in the structure of oxide carriers. IY. Macro- and microdistribution of $MoS_2$ in $\gamma$-$Al_2O_3$ pores // Kinetics and Catalysis. T. 33. №1, 1992, -C. 170-175.

**160.** Startsev A.N. Sulfide hydrotreating catalysts: synthesis, structure, properties. - Novosibirsk: 2007. 203.

**161.** Sidelkovskaya V.G., Surin S.A. et al. Spectroscopic study of the interaction of components in $NiMoO_4$-based aluminonickelmolybdenum catalysts // Kinetics and Catalysis - Moscow. 1988. T.29. - №6. -C.1513-1517.

**162.** Agievsky D.A., Kvashonkin V.I., Zadko I.I. et al. Influence of the content of Ni-Mo components in hydrotreating catalysts on their activity and interaction with alumina carrier // Kinetics and Catalysis 1984. vol.25. vol.1. -C.178-185.
**163.** Agievsky D.A., Kvashonkin V.I. et al. Influence of Aluminum Nitrate on Porous Carrier Structure and Phase Composition of Aluminonickel-Molybdenum Catalysts // Kinetics and Catalysis 1984. vol. 25. vol. 4. - C.928-933.
**164.** Lurie MA, Kurtz I.Z., Storozheva L.N. et al. Hydrotreating of heavy oil feedstock on catalysts with different porous structure // KK. 1991. V.32. vol. 6.- P.1399-1405.
**165.** Fouad K., Malika B., Mathieu D., Pierre A.. Control of the textural properties of nanocrystalline boehmite (γ-AlOOH) regarding its peptization ability // Powder Technology, 2013,-#237. - P. 602-609.
**166.** Deutschmann O., Knozinger H., Kochloefl K., Turek T. Heterogeneous Catalysis and Solid Catalysts // 2009.-P. 110.
**167.** Ismagilov Z.R., Shkrabina R.A., Koryabkina N.A. Alumina-oxide carriers: production, properties and application in catalytic processes of environmental protection // Novosibirsk 1998. - C. 82.
**168.** Smolikov, M.D.; Belyi, A.S.; Udras, I.E.; Kiryanov, D.I. Design of active component distribution over pores in catalysts of hydrocatalytic processes of oil refining (in Russian) // Russian Chemical Journal of D.I.Mendeleev Society. 2007, vol. LI, No. 4. -C. 48-56.
**169.** Li D.-Y., Lin Y.-S., Li Y.-C., Shieh D.-L., & Lin J.-L. Synthesis of mesoporous pseudoboehmite and alumina templated with 1-hexadecyl-2,3-dimethyl-imidazolium

chloride // Microporous and Mesoporous Materials, 2008, - #108. - P. 276-282.

**170.** Huang B., Bartholomew C. H., Smith S. J., Woodfield, B. F. Facile solvent-deficient synthesis of mesoporous γ-alumina with controlled pore structures // Microporous and Mesoporous Materials, 2013.- #165.-P. 70-78.

**171.** Kim Y., Kim C., Kim P., Yi J. Effect of preparation conditions on the phase transformation of mesoporous alumina // Journal of Non - Crystalline Solids, 2005. -№ 351.- P. 550 - 556.

**172.** Huirache-Acuna R., T.A. Zepeda E.M. Rivera-Munoz R. Nava C.V. Loricera B. Pawelec. Characterization and HDS performance of sulfide CoMoW catalysts supported on mesoporous Al-SBA-16 substrates // Fuel. 2014. Journal homepage: www. Elsevier. Com/locate/fuel. - P. 1-14.

**173.** X. Li, Han D., Xu Y., Liu X., Yan Z. Bimodal mesoporous γ-Al2O3: A promising support for CoMo - based catalyst in hydrodesulfurization of 4,6-DMDBT // Materials Letters. - 2011. - V.65. - P.1765-1767.

**174.** Catalyst, method of its preparation and process of hydrodessulfurization of diesel fractions. Yashnik S.A., Ismailov Z.R., Surovtsev T.A., Noskov A.S., Bukhtiyarova G.A. // Patent RU № 2314154. - 10.01.2008г. Bul. No. 1.

**175.** Method of preparation of phosphorus-containing aluminum cobaltmolybdenum or aluminum nickelmolybdenum catalyst for hydrotreatment of hydrocarbon feedstock. Vishnitsky A.B., Levchenko A. L., Fedorov V.I., Morokhovsky B.K., Wildt N.G., Pavlova T.V., Bayrak N.I., Mokray V., Manusova L. // Patent RU № 2107546. - 27.03.1998г.

**176.** Method of preparation of catalyst for hydrotreating of petroleum products. Shebanov S.M., Shipkov N.N., Strelkov V.A.//Pat. RU №2100079. -27.12.1997г.
**177.** A catalyst, a method of its preparation, a method of obtaining a carrier for this catalyst and a process of hydrodesulfurization of diesel fractions. Yashnik S.A., Ismailov Z.R., Surovtsev T.A., Noskov A.S., Bukhtiyarova G.A. // Patent RU № 2313389. - 27.12.2007г. Bulletin No. 36.
**178.** Catalyst, method of its preparation and method of refining of diesel distillates. Bukhtiyarova GA, Nuzhdin AL, Aleshina GI, Noskov AS // Patent RU № 2468864 - 10.12.2012g. Bulletin No. 34.
**179.** Hydrotreating catalyst for hydrocarbon feedstock, carrier for hydrotreating catalyst, method of preparation of carrier, method of preparation of catalyst and method of hydrotreating of hydrocarbon feedstock. Klimov O.V., Koryakina G.I., Leonova K.A., Budukva S.V., Pereyma V.Yu., Dick P.P., Noskov A.S., Parakhin O.A. // Patent RU №2478428 - 10.04.2013. Bulletin No. 10
**180.** Catalyst for deep hydrotreating of oil fractions, method of its preparation. Pimerzin AA, Tomina NN, Maximov NM, Solmanov PS // Patent RU № 2497586.- 10.11.2013.
**181.** Pashigreva A.V., Bukhtiyarova G.A., Klimov O.V., Litvak G.S., Noskov A.S. Effect of heat treatment conditions on the activity of the catalyst for deep hydrotreating of diesel fractions $CoMo/Al_2O_3$ // Kinetics and Catalysis, 2008. vol. 49, -№ 6, -C. 855-864.
**182.** A catalyst, a method for preparing a carrier, a method for preparing a catalyst and a method for hydrotreating hydrocarbon feedstock. Klimov O. V., Koryakina G.I.,

Budukva S. V., Leonova K.A., Pereima V.Y., Dik P.P., Noskov A.S., Parakhin O. A. // Patent RU №2472585. - 20.01.2013. Bulletin No. 2.
**183.** Method of preparation of catalyst for hydrotreating of oil fractions. Aliev R.R., Porublev M.A., Zelentsov Y.N., Tselutina M.I., Yaskin V.P., Elshin A.I., Osokina N.A. Patent RU 2084285 - 20.07.1997.
**184.** Catalyst for hydrotreating of oil feedstock and method of its preparation. Vyazkov V.A., Levin O.V., Vlasov V.G., Loginova A.N., Tomina N.N., Sharikhina M.A., Shafransky E.L., Lyadin N.M., Borisov V.P., Oltyrev A.G. Patent RU 2137541 - 20.09.1999.
**185.** Chen W., Maugé F., van Gestel J., Nie H., Li D., Long X. Effect of modification of the alumina acidity on the properties of supported Mo and CoMo sulfide catalysts // Journal of Catalysis. -2013. -V.304. -P.47-62.
**186.** Rashidi F., Sasaki T., Rashidi A.M., Kharat A.N., Jozani K.J.. Ultra deep hydrodesulfurization of diesel fuels using highly efficient nanoalumina-supported catalysts: Impact of support, phosphorus, and/or boron on the structure and catalytic activity // Journal of Catalysis. -2013. -V.299. -P.321-335.
**187.** Ferdous D., Dalai A.K., Adjaye J. A series of NiMo/$Al_2O_3$catalysts containing boron and phosphorus. Part II. Hydrodenitrogenation and hydrodesulfurization using heavy gasoil derived from Athabasca bitumen // Applied Catalysis A: General, 2004. -V.260. -P.153-162.
**188.** Tomina N.N., Pimerzin A.A., Moiseev I. K. Sulfide catalysts for hydrotreating of oil fractions // Ros Khim. zh. (Zh. Ros. Khim. obs. named after D.I. Mendeleev), 2008, vol. III, - № 4. - C. 41-52.

**189.** Topsøe H., Clausen B.S.. Importance of Co-Mo-S type structures in hydrodesulphurization // Catal. Rev.-Sci. Eng. 1984.-№26 (3-4), -P. 395-420.
**190.** Clausen B.S., Lengeler B., Topsøe H.. X-ray absorption spectroscopy studies of calcined Mo-$Al_2O_3$ and Co-Mo-$Al_2O_3$ hydrodesulphurization catalysts // Polyhedron1986. -№5(1-2) -P.199-202.
**191.** Eisbouts S. On the flexibility of active phase in hydrotreating catalysts // Appl. Catal., 1997. -V.158. -P.53-92.
**192.** Cattenot M., Geantet, C., Glasson C., M. Breysse. Promoting effect of ruthenium on NiMo/Al2O3 hydrotreating catalysts // Appl. Catal., 2001. -V.213. - P. 217-224.
**193.** Topsøe H. The role of Co-Mo-S type structures in hydrotreating catalysts // Appl. Catal.,2007. - V.322. - P.3-8.
**194.** Brorson M., Carlsson A., Topsøe H. The morphology of MoS2, WS2, Co-Mo-S, Ni-Mo-S and Ni-W-S nanoclusters in hydrodesulfurization catalysts revealed by HAADF-STEM // Catal. Today.,2007. -V.123. -P.31-36.
**195.** Besenbacher F., Brorson M., Clausen B.S., Helveg S., Hinnemann B., Kibsgaard J., Lauritsen J.V., Moses P.G., Norskov J.K., Topsoe H. Recent STM, DFT and HAADF-STEM studies of sulfide-based hydrotreating catalysts: Insight into mechanistic, structural and particle size effects // Catal. Today., 2008. -V.130. - P.86-96.
**196.** Kibsgaard J., Tuxen A., Knudsen K. G., Brorson M., Topsoe H., Laegsgaard E., Lauritsen J.V., Besenbacher F. Comparative atomic-scale analysis of promotional effects by late 3d-transition metals in $MoS_2$ hydrotreating catalysts // J. Catal. Catal., 2010. - V. 272. - P. 195-203.

**197.** Lauritsen J.V., Bollinger M.V., Lægsgaard E., Jacobsen K.W., Norskov J.K., Clausen B.S., Topsoe H., Besenbacher F. Atomic-scale insight into structure and morphology changes of $MoS_2$ nanoclusters in hydrotreating catalysts // J. Catal. Catal., 2004. -V.221. -P.510-522.
**198.** Kogan V.M., Parfenova N.M., Gaziev R.G. et al. In situ radioisotopic study of active centers of sulfide Co-Mo catalysts and mechanism of thiophene hydrodesulfurization // Kinetics and Catalysis, 2003. T. 44. -№ 4,.- C.638-656.
**199.** Tuxen A., Gøbel H., Hinnemann B., Li Z., Knudsen K.G, Topsoe H., Lauritsen J.V., Besenbacher F. An atomic-scale investigation of carbon in $MoS_2$ hydrotreating catalysts sulfide by organosulfur compounds // Journal of Catalysis, 2011. -V.281. -P.345-351.
**200.** Liu B., Chai Y., Li Y., Wang A., Liu Y., Liu C.. Effect of sulfidation atmosphere on the performance of the CoMo/$\gamma$-$Al_2O_3$ catalysts in hydrodesulfurization of FCC gasoline // Applied Catalysis A: General. 2014. -V.471. -P.70-79
**201.** Pashigreva A.V., Bukhtiyarova G.A., Kashkin V.N., Ivanova A.S., Kulko E.V., Noskov A.S., Activity of Co-Mo catalysts in the transformation of individual sulfur-containing compounds // Proc. of XIII Mendeleev Congress on General and Applied Chemistry, September 23-28, 2007. Moscow, 2007. -T. 3. -C. 419.
**202.** Mashkina A.V., Sakhaltueva L.G. Gas-phase hydrogenation of thiophene into tetrahydrothiophene in the presence of sulfide catalysts // Kinetics and Catalysis. 2002. T.43. -№1. -C.116-124.
**203.** Mohammed A.N. Hydrodesulfurization of thiophene over $CoMo/Al_2O_3$ catalyst using fixed- and fluidized-bed

reactors // Journal of Engineering. February 2011 - No. 1 V. 17. -P.92-102.

**204.** Ghanbari K.*, Mohammadi M. and Tajerian M.. The effect of mass transfer resistance on the kinetics of thiophene hydrodesulfurization// Petroleum & Coal 2006.-#48 (2). -P.33-36.

**205.** Hamid A. Al-Megren* Hydrodesulfurization of thiophene over bimetallic ni-mo sulfide catalysts prepared by different methods // The Arabian Journal for Science and Engineering, January 2009.- V.34.- No. 1A. -P. 55-66.

**206.** Nino R., Masahiro Y., Takeshi K., and Yasuaki O. Hydrodesulfurization Activity of Co-Mo/$Al_2O_3$ // Catalysts Prepared with Citric Acid: Post-treatment of Calcined Catalysts with High Mo Loading Journal, 2010.-No. 53, (5). -P. 292-302 .

**207.** Leonidas E. Kallinikos A, Andreas J., Nikos G. Kinetic study and $H_2S$ effect on refractory DBTs desulfurization in a heavy gasoil // Journal of Catalysis, 2010.-# 269. -P. 169-178.

**208.** Muhammada Y., Yingzhou Lu, Chong Sh., Chunxi Li. Dibenzothiophene hydrodesulfurization over Ru promoted alumina-based catalysts using in situ generated hydrogen // Energy Conversion and Management, 2011.No.52-P.1364-1370.

**209.** Malailak N., Sirisuk A. Hydrodesulfurization of dibenzothiophene over como catalysts on $Al_2O_3$-tio2 mixed oxide supports // Pure and applied chemistry international conference, 2012.- P. 417-420.

**210.** Isayeva N.F., Mirzaeva E.I., Paisiev J., Nasullaev H., Abdurakhimov M.U. Effect of the method of preparation on the porous characteristics of carriers for forecontacts and catalysts for hydrotreatment of oils // Actual problems of oil and gas purification from impurities by various physical and chemical methods: Proceedings of the Republican Scientific and Technical Conference, May 20-21, 2011. - Karshi, 2011. - C. 74-75.

**211.** Gulyamov Sh.T., Egamberdiev A., Yakhyaeva R., Isayeva N.F. Synthesis of carriers for catalysts for hydrotreating of oil distillates // //.

Umidli kimyogarlar - 2010: Proceedings of the Scientific and Technical Conference of Young Scientists: doctoral students, postgraduates, researchers and students of undergraduate and graduate programs. April 6-9, 2010. - Tashkent, 2010. VOL.1. - P.122-123.

**212.** Isayeva N.F., Jalalova Sh.B. Technology of preparation of carriers based on aluminum hydroxide of different shelf life // Uzbek Journal of Oil and Gas, Tashkent, 2012. - №1, - C.27-30.

**213.** Yunusov M.P., Molodozhenyuk T.B., Ergashev M.M. et al. Investigation of the protective layer system for hydrotreating of oil distillates from Uzbek oils // Chemical Industry, St. Petersburg. 2007. - №2. - C. 55-61.

**214.** Dmitrieva N.N., Gashenko G.A., Isayeva N.F. Synthesis and properties of zeolite-containing carriers for oil hydrotreating catalysts // Oil and gas sanoati kimoviy tekhnologilarinini dolzarb muammolari: Uzbekiston republican oliy v orta makhsus tahsulim vazirligi mikyosidagi ilmiy-amaliy conference. April 24-25, 2009. - Karshi, 2009. - C. 83.

**215.** Gashenko GA, Isaeva NF, Nasullaev HA, Lavoshnikov VV, Jalalova SB, Molodozhenyuk TB, Yunusov MP Obtaining cation-exchange forms of zeolitization products of

Angren kaolins for modification of carriers and catalysts // Journal of Chemical Industry, St. Petersburg, 2011. - № 5. - C. 223-231.
**216.** Jalalova S.B., Nasullaev H., Gulomov Sh., Isayeva N.F., Teshabaev Z.A., Saidullaev B.T. Properties of carriers forecontact-catalyst protective layer using waste production and local mineral raw materials // Actual problems of oil and gas purification from impurities by various physical and chemical methods: Proceedings of the Republican Scientific and Technical Conference. May 20-21, 2011. - Karshi, 2011. - C. 70-72.
**217.** Yunusov MP, Jalalova S.B., Nasullaev H.A., Gulyamov Sh.T, Mirzaeva E.I., Isaeva N.F., Molodozhenyuk T.B. Effect of the method of application of active components on the distribution of hydrating metals in hydrotreating catalysts // Chemical Industry, 2013.- Vol.90.- No.3. -C.141- 150.
**218.** Yunusov M.P., Saidaxmedov Sh.M., Djalalova Sh.B., Nasullaev Kh.A., Gulyamov Sh.T., Isaeva N.F., Mirzaeva E.I. Synthesis and research of Co - Ni-Mo catalysts of oil fractions hydroprocessing // Catalysis for sustainable energy. Poland. 2015.-№ 2. -P. 43-56.
**219.** Ergashev M.M. Development of a special layer catalyst for oil hydrotreating unit and organization of its production: Cand.Sci. - Tashkent: UzKFITI, 2008. - 118c.
**220.** Yunusov M.P., Isayeva N.F., Teshabaev Z.A., Musayeva G.H., Abdurakhimov M.U., Balaev V. Multilayer catalyst systems one of the ways to improve the efficiency of oil production // Actual problems of oil and gas purification from impurities by various physical and chemical methods:

Proceedings of the Republican Scientific and Technical Conference. May 20-21, 2011. - Karshi, 2011. - C. 65-67.
**221.** Isayeva N.F., Jalalova Sh.B., Teshabaev Z.A., Yunusov M.P. Synthesis of samples of carriers of catalysts for hydrotreating oils using local mineral additive // Actual problems of oil and gas processing in Uzbekistan: Proceedings of the Republican Scientific and Technical Conference. October 7-8, 2009. - Bukhara, 2009. - C. 195-197
**222.** Yunusov MP, Jalalova Sh.B., Molodozhenyuk T.B., IsaevaN.F., Mirzaeva E.I., Gulomov Sh.T., Mahkamov H.M., Bahramov R., Nasullaev H.A. Catalyst for hydrotreatment of oil fractions and method of preparation // / / Intellectual Property Agency of the Republic of Uzbekistan, №IAP № 05423. - 01.06.2017г.
**223.** Kuzmicheva E.L., Mahkamov H.M., Molodozhenyuk T.B. Effect of modifying additives on the surface and catalytic properties of platinum and palladium catalysts // Chemical Industry, 2001, - № 11, - P.10-16.
**224.** Yunusov M.P., Djalalova Sh.B., Isaeva N.F., Molodojenyuk T.B., Teshabaev Z.A. Working out of Catalyst Systems for Hydrodesulphurization of Oil Fractions in Uzbekistan // Catalysis in solving problems of petrochemistry and oil refining: Proceedings of Azerbaijan-Russian Symposium. September 28-30, 2010. - Baku 2010. - C. 71-72.
**225.** Nasullaev H.A., Egamberdiev AP, Faiziev J.B., Isayeva N.F. Study of acid-base properties in the synthesis of alumina carriers // Bulletin of NUUz. - Tashkent, 2010. - № 4. - C. 107-109.

**226.** Isayeva N.F., Nasullaev H.A., Usmanov R.M., Khudojberdiyeva U., Abdurakhimov M.U. Synthesis of carriers for catalysts of hydroprocesses // Bulletin of NUUz. - Tashkent, 2012. - № 3/1. - C. 136-139.
**227.** Jalalova Sh.B., H.A. Nasullaev, Isaeva N.F., Gulomov Sh.T., M.P. Yunusov. Genesis of redox centers on the surface of hydroprocess catalysts during preparation and activation // Chemical Industry, 2015. -T.92.-№2. -C. 94-97
**228.** Method of preparation of catalyst for hydrotreating of oil fractions Yolshin A.I., Aliev A.I., Porublev M.A., Osokina N.A., Tselutina M.I. Kuks I.V. // Pat. RF., № 2206396 -20.06.2003.
**229.** Lulic P. Influence of catalysis on optimal commercial refining // Erdol Erdgas Kohle, 2001. -№117. N 12. -P. 583-585.
**230.** Alexandrov P.V., Kashkin V.N., Bukhtiyarova GA, Nuzhdin AL, Pashigreva AV, Klimov OV, Noskov AS Comparative study of $NiMo/Al_2O_3$ and $CoMo/Al_2O_3$ catalysts of a new generation in the reactions of hydrodesulfurization and hydrodeazotization of vacuum gas oil // Oil refining and petrochemistry 2010. Novosibirsk № 9 - P.3-9.
**231.** Patrie W., Turner W.J.. Zeuthen. New trimetallic catalyst improves FCCU // Oil and Gas Journal, 2001.-#99. N 23.- P.56-60.
**232.** Gaya U.I, Jibril B.Y., Al-Wehaibi Y.M., Al-Hajri R.S. and Naser J.T.. Ring Opening of Decalin over Zeolite-Supported Ni-Co-Mo Catalysts // International Conference on Environment, Chemistry and Biology IPCBEE 2012. - Vol.49.-P.139-143.

**233.** Arul D., Arash E., Kenneth S. Sonochemical Preparation of Supported Hydrodesulfurization Catalysts. //J. Am. Chem. Soc. 2001, -№123. -P.8310-8316
**234.** Isayeva N.F. Study of oxide precursors of active structures in AI-Ni-Mo, Al-Co-Mo and Al-Co-Ni-Mo // Uzbek Chemical Journal. 2011, Special issue. - C. 126-128.
**235.** Isayeva N.F., Amanov N.G., Molodozhenyuk T.B., Yunusov M.P. "Study of the influence of the conditions of preparation of crust type catalysts on the thickness of the crust" // Technologies of processing of local raw materials and products: Proceedings of the Republican Scientific - Technical Conference - Tashkent, 2009. - C. 110-111
**236.** Isayeva N.F., Yunusov M.P. Development of an effective catalyst for hydrodehydration of oils. Proceedings of the republican scientific and technical conference. "Actual problems of innovative technologies of chemical, oil and gas and food industries" // Kungrad - 2010, October 28-29, - P.12-13.
**237.** Isayeva N.F., Thermographic study of the synthesis of trimetallic catalysts for hydrotreating of oils // Proceedings of the scientific and technical conference. Actual problems of innovative technologies of chemical, oil and gas and food industries. Tashkent. November 22, 2012.-C. 122-123.
**238.** Yunusov MP, Jalalova Sh.B, Nasullaev H.A , Isaeva N.F., Mirzaeva E.I., Saidakhmedov S.M. Effect of modification by $Zn^{2+}$ ions on the catalytic properties of hydrotreating and hydrodearomatization catalysts // Uzbek Journal of Oil and Gas. 2014, -№4. -C. 41-44.
**239.** Isaeva N.F. The innovative synthesis trimetallic oil hydrotreating catalysts // Innovation for sustainability-harmonizing science, technology and economic development

with human and natural environment. "Proceedings of the uzbek-japan symposium on ecotechnologies". 14 may 2016 - Tashkent.- C. 48-53.
**240.** Isayeva N.F. Effect of citric acid on the properties of catalysts hydrotreating oil fractions // Proceedings of the Republican Scientific and Practical Conference "Actual problems of chemical science and innovative technologies of its training. Tashkent 2016. March 30-31. - C.103-104.
**241.** Isaeva N. Phase composition of carriers and trimetallic catalysts synthesized with the participation of transition metal chelate complexes // Proceedings of the Republican Scientific and Technical Conference "Oil and gas processing, alternative fuels" Tashkent-2016. -C.124-126.
**242.** Isaeva N.F. Development of a new catalyst hydrodesulfurization of petroleum oil fractions // II International Scientific and Technical Conference "Innovative developments in the field of chemistry and technology of fuels and lubricants" October 19-20, 2017. - Bukhara, 2017, -C. 188-190
**243.** Jalalova Sh.B., Isayeva N.F., Nasullaev H.A., Abdurakhimov M.U. Study of processes occurring during the synthesis of catalyst AKNM-3/5 on the example of model systems // Bulletin of NUUz., 2012, - № 3/1. - C. 192-195.
**244.** Isayeva N.F., Yunusov M.P., Molodozhenyuk T.B., Toshtemirov B.V., Nasullaev H.A. New applied trimetallic catalyst for hydrodehydration of oil fractions of oil // Uzbek Journal of Oil and Gas, Tashkent -2011. -№2. -C. 39-40
**245.** Landau MV, Alekseenko LN, Vinogradova OV, Mikhailov VI, Nefedov BK Influence of deposition conditions and heat treatment of nickel molybdate on the

composition and structure of the resulting product // Kinetics and Catalysis, 1989 .- Vol.30. vol.4. - P.933-938.

**246.** Yunusov M.P., Molodozhenyuk T.B., Jalalova Sh.B., Isayeva N.F., Gulomov Sh., Nasullaev H., Mirzaeva E.N. Multilayer catalytic sistems as one of the ways to increase efficiency of fuel and oil hydrofining // 21th European Congress on catalysis "20 years of European catalysis... and beyond" 1-6 september 2013 Lyon, France. 21st European Congress on catalysis "20 years of European Catalysis... and beyond" 1-6 september 2013 Lyon, France.

**247.** Isayeva N.F., Yunusov M.P., Teshabaev Z.A. Effect of the method and conditions of formation of cobalt-nickel-molybdenum catalyst for hydrotreating of oils // ROSKATALIZ: Proceedings of the Russian Congress on Catalysis. October 3-7, 2011. Moscow. - Novosibirsk, 2011. VOL. II. - C. 46.

**248.** Yunusov M.P. Jalalova S.B., Isayeva N.F., Mahkamov H.M. Problems of synthesis of carriers for titanium containing hydrotreating catalysts by molecular layering // Symposium "Modern Problems of Nanocatalysis" September 24-28, 2012. -Uzhgorod -2012. -C75-76.

**249.** Yunusov M.P., Jalalova Sh.B., Isayeva N.F., Development and implementation of multilayer catalyst characteristics of fuels and oils // International Conference "catalytic processes of oil refining, petrochemistry and ecology" October 14-16, 2013. - Tashkent. Abstracts Novosibirsk - 2013. - C. 11-12.

**250.** Yunusov M.P., Isayeva N.F., Amanov N.G., Molodozhenyuk T.B. Synthesis and properties of elements of multilayer catalytic systems for hydrotreating reactors //

Eurasian Symposium on Innovations in Catalysis and Electrochemistry: May 26-28, 2010. - Almaty, 2010. 40.
**251.** Yunusov MP, Jalalova Sh.B., Mirzaeva E.I., Gulomov Sh.T., Nasullaev H.A. Isaeva N.F., Yunusov Analysis of experience in industrial operation of multilayer catalytic systems // "Rosscatalysis 2014" II Russian Congress on Catalysis. Collection of abstracts. 2 vol. October 2-5, 2014. Samara. -Novosibirsk, 2014. C.314.

Printed by Books on Demand GmbH, Norderstedt / Germany